ELEMENTS OF CHEMISTRY,

BY MR LAVOISIER,

TRANSLATED BY ROBERT KERR

ADVERTISEMENT OF THE TRANSLATOR.

The very high character of Mr Lavoisier as a chemical philosopher, and the great revolution which, in the opinion of many excellent chemists, he has effected in the theory of chemistry, has long made it much desired to have a connected account of his discoveries, and of the new theory he has founded upon the modern experiments written by himself. This is now accomplished by the publication of his Elements of Chemistry; therefore no excuse can be at all necessary for giving the following work to the public in an English dress; and the only hesitation of the Translator is with regard to his own abilities for the task. He is most ready to confess, that his knowledge of the composition [Pg vi]of language fit for publication is far inferior to his attachment to the subject, and to his desire of appearing decently before the judgment of the world.

He has earnestly endeavoured to give the meaning of the Author with the most scrupulous fidelity, having paid infinitely greater attention to accuracy of translation than to elegance of stile. This last indeed, had he even, by proper labour, been capable of attaining, he has been obliged, for very obvious reasons, to neglect, far more than accorded with his wishes. The French copy did not reach his hands before the middle of September; and it was judged necessary by the Publisher that the Translation should be ready by the commencement of the University Session at the end of October.

He at first intended to have changed all the weights and measures used by Mr Lavoisier into their correspondent English denominations, but, upon trial, the task was found infinitely too great for the time allowed; and to have executed this part of the work inaccurately, must have been both useless and misleading to the reader. All that has been attempted in this way is adding, between brackets ( ), the degrees of Fahrenheit's[Pg vii] scale corresponding with those of Reaumeur's thermometer, which is used by the Author. Rules are added, however, in the Appendix, for converting the French weights and measures into English, by which means the reader may at any time calculate such quantities as occur, when desirous of comparing Mr Lavoisier's experiments with those of British authors.

By an oversight, the first part of the translation went to press without any distinction being preserved between charcoal and its simple elementary part, which enters into chemical combinations, especially with oxygen or the acidifying principle, forming carbonic acid. This pure element, which exists in great plenty in well made charcoal, is named by Mr Lavoisier *carbone*, and ought to have been so in the translation; but the attentive reader can very easily rectify the mistake. There is an error in Plate XI. which the engraver copied strictly from the original, and which was not discovered until the plate was worked off at press, when that part of the Elements which treats of the apparatus there represented came to be translated. The two tubes 21. and 24. by which the gas is conveyed[Pg viii] into the bottles of alkaline solution 22. 25. should have been made to dip into the liquor, while the other tubes 23. and 26. which carry off the gas, ought to have been cut off some way above the surface of the liquor in the bottles.

A few explanatory notes are added; and indeed, from the perspicuity of the Author, very few were found necessary. In a very small number of places, the liberty has been taken of throwing to the bottom of the page, in notes, some parenthetical expressions, only relative to the subject, which, in their original place, tended to confuse the sense. These, and the original notes of the Author, are distinguished by the letter A, and to the few which the Translator has ventured to add, the letter E is subjoined.

Mr Lavoisier has added, in an Appendix, several very useful Tables for facilitating the calculations now necessary in the advanced state of modern chemistry, wherein the most scrupulous accuracy is required. It is proper to give some account of these, and of the reasons for omitting several of them.[Pg ix]

No. I. of the French Appendix is a Table for converting ounces, gros, and grains, into the decimal fractions of the French pound; and No. II. for reducing these decimal fractions again into the vulgar subdivisions. No. III. contains the number of French cubical inches and decimals which correspond to a determinate weight of water.

The Translator would most readily have converted these Tables into English weights and measures; but the necessary calculations must have occupied a great deal more time than could have been spared in the period limited for publication. They are therefore omitted, as altogether useless, in their present state, to the British chemist.

No. IV. is a Table for converting lines or twelfth parts of the inch, and twelfth parts of lines, into decimal fractions, chiefly for the purpose of making the necessary corrections upon the quantities of gasses according to their barometrical pressure. This can hardly be at all useful or necessary, as the barometers used in Britain are graduated in decimal fractions of the inch, but, being referred to by the Author in[Pg x] the text, it has been retained, and is No. I. of the Appendix to this Translation.

No. V. Is a Table for converting the observed heights of water within the jars used in pneumato-chemical experiments into correspondent heights of mercury for correcting the volume of gasses. This, in Mr Lavoisier's Work, is expressed for the water in lines, and for the mercury in decimals of the inch, and consequently, for the reasons given respecting the Fourth Table, must have been of no use. The Translator has therefore calculated a Table for this correction, in which the water is expressed in decimals, as well as the mercury. This Table is No. II. of the English Appendix.

No. VI. contains the number of French cubical inches and decimals contained in the corresponding ounce-measures used in the experiments of our celebrated countryman Dr Priestley. This Table, which forms No. III. of the English Appendix, is retained, with the addition of a column, in which the corresponding English cubical inches and decimals are expressed.[Pg xi]

No. VII. Is a Table of the weights of a cubical foot and inch, French measure, of the different gasses expressed in French ounces, gros, grains, and decimals. This, which forms No. VI. of the English Appendix, has been, with considerable labour, calculated into English weight and measure.

No. VIII. Gives the specific gravities of a great number of bodies, with columns, containing the weights of a cubical foot and inch, French measure, of all the substances. The specific gravities of this Table, which is No. VII. of the English Appendix, are retained, but the additional columns, as useless to the British philosopher, are omitted; and to have converted these into English denominations must have required very long and painful calculations.

Rules are subjoined, in the Appendix to this translation, for converting all the weights and measures used by Mr Lavoisier into corresponding English denominations; and the Translator is proud to acknowledge his obligation to the learned Professor of Natural Philosophy in the University of Edinburgh, who kindly supplied him with the necessary information for this purpose. A Table is likewise added, No. IV. of[Pg xii] the English Appendix, for converting the degrees of Reaumeur's scale used by Mr Lavoisier into the corresponding degrees of Fahrenheit, which is universally employed in Britain[1].

This Translation is sent into the world with the utmost diffidence, tempered, however, with this consolation, that, though it must fall greatly short of the elegance, or even propriety of language, which every writer ought to endeavour to attain, it cannot fail of advancing the interests of true chemical science, by disseminating the accurate mode of analysis adopted by its justly celebrated Author. Should the public call for a second edition, every care shall be taken to correct the forced imperfections of the present translation, and to improve the work by valuable additional matter from other authors of reputation in the several subjects treated of.

EDINBURGH, }
Oct. 23. 1789. }

FOOTNOTES:

[1]The Translator has since been enabled, by the kind assistance of the gentleman above alluded to, to give Tables, of the same nature with those of Mr Lavoisier, for facilitating the calculations of the results of chemical experiments.

[Pg xiii]

PREFACE OF THE AUTHOR.

When I began the following Work, my only object was to extend and explain more fully the Memoir which I read at the public meeting of the Academy of Sciences in the month of April 1787, on the necessity of reforming and completing the Nomenclature of Chemistry. While engaged in this employment, I perceived, better than I had ever done before, the justice of the following maxims of the Abbé de Condillac, in his System of Logic, and some other of his works.

"We think only through the medium of words.—Languages are true analytical methods.—Algebra,[Pg xiv] which is adapted to its purpose in every species of expression, in the most simple, most exact, and best manner possible, is at the same time a language and an analytical method.—The art of reasoning is nothing more than a language well arranged."

Thus, while I thought myself employed only in forming a Nomenclature, and while I proposed to myself nothing more than to improve the chemical language, my work transformed itself by degrees, without my being able to prevent it, into a treatise upon the Elements of Chemistry.

The impossibility of separating the nomenclature of a science from the science itself, is owing to this, that every branch of physical science must consist of three things; the series of facts which are the objects of the science, the ideas which represent these facts, and the words by which these ideas are expressed. Like three impressions of the same seal, the word ought to produce the idea, and the idea to be a picture of the fact. And, as ideas are preserved and communicated by means of words, it necessarily follows[Pg xv] that we cannot improve the language of any science without at the same time improving the science itself; neither can we, on the other hand, improve a science, without improving the language or nomenclature which belongs to it. However certain the facts of any science may be, and, however just the ideas we may have formed of these facts, we can only communicate false impressions to others, while we want words by which these may be properly expressed.

To those who will consider it with attention, the first part of this treatise will afford frequent proofs of the truth of the above observations. But as, in the conduct of my work, I have been obliged to observe an order of arrangement essentially differing from what has been adopted in any other chemical work yet published, it is proper that I should explain the motives which have led me to do so.

It is a maxim universally admitted in geometry, and indeed in every branch of knowledge, that, in the progress of investigation, we should proceed from known facts to what is unknown. In early infancy, our ideas spring from our wants; the sensation of want excites the idea of[Pg xvi] the object by which it is to be gratified. In this manner, from a series of sensations, observations, and analyses, a successive train of ideas arises, so linked together, that an attentive observer may trace back to a certain point the order and connection of the whole sum of human knowledge.

When we begin the study of any science, we are in a situation, respecting that science, similar to that of children; and the course by which we have to advance is precisely the same which Nature follows in the formation of their ideas. In a child, the idea is merely an effect produced by a sensation; and, in the same manner, in commencing the study of a physical science, we ought to form no idea but what is a necessary consequence, and immediate effect, of an experiment or observation. Besides, he that enters upon the career of science, is in a less advantageous situation than a child who is acquiring his first ideas. To the child, Nature gives various means of rectifying any mistakes he may commit respecting the salutary or hurtful qualities of the objects which surround him. On every occasion his judgments are corrected by experience; want and pain are the necessary[Pg xvii] consequences arising from false judgment; gratification and pleasure are produced by judging aright. Under such masters, we cannot fail to become well informed; and we soon learn to reason justly, when want and pain are the necessary consequences of a contrary conduct.

In the study and practice of the sciences it is quite different; the false judgments we form neither affect our existence nor our welfare; and we are not forced by any physical necessity to correct them. Imagination, on the contrary, which is ever wandering beyond the bounds of truth, joined to self-love and that self-confidence we are so apt to indulge, prompt us to draw conclusions which are not immediately derived from facts; so that we become in some measure interested in deceiving ourselves. Hence it is by no means to be wondered, that, in the science of physics in general, men have often made suppositions, instead of forming conclusions. These suppositions, handed down from one age to another, acquire additional weight from the authorities by which they are supported, till at last they are received, even by men of genius, as fundamental truths.[Pg xviii]

The only method of preventing such errors from taking place, and of correcting them when formed, is to restrain and simplify our reasoning as much as possible. This depends entirely upon ourselves, and the neglect of it is the only source of our mistakes. We must trust to nothing but facts: These are presented to us by Nature, and cannot deceive. We ought, in every instance, to submit our reasoning to the test of experiment, and never to search for truth but by the natural road of experiment and observation. Thus mathematicians obtain the solution of a problem by the mere arrangement of data, and by reducing their reasoning to such simple steps, to conclusions so very obvious, as never to lose sight of the evidence which guides them.

Thoroughly convinced of these truths, I have imposed upon myself, as a law, never to advance but from what is known to what is unknown; never to form any conclusion which is not an immediate consequence necessarily flowing from observation and experiment; and always to arrange the facts, and the conclusions which are drawn from them, in such an order as shall render it most easy for beginners in the[Pg xix] study of chemistry thoroughly to understand them. Hence I have been obliged to depart from the usual order of courses of lectures and of treatises upon chemistry, which always assume the first principles of the science, as known, when the pupil or the reader should never be supposed to know them till they have been explained in subsequent lessons. In almost every instance, these begin by treating of the elements of matter, and by explaining the table of affinities, without considering, that, in so doing, they must bring the principal phenomena of chemistry into view at the very outset: They make use of terms which have not been defined, and suppose the science to be understood by the very persons they are only beginning to teach. It ought likewise to be considered, that very little of chemistry can be learned in a first course, which is hardly sufficient to make the language of the science familiar to the ears, or the apparatus familiar to the eyes. It is almost impossible to become a chemist in less than three or four years of constant application.

These inconveniencies are occasioned not so much by the nature of the subject, as by the method of teaching it; and, to avoid them, I[Pg xx] was chiefly induced to adopt a new arrangement of chemistry, which appeared to me more consonant to the order of Nature. I acknowledge, however, that in thus endeavouring to avoid difficulties of one kind, I have found myself involved in others of a different species, some of which I have not been able to remove; but I am persuaded, that such as remain do not arise from the nature of the order I have adopted, but are rather consequences of the imperfection under which chemistry still labours. This science still has many chasms, which interrupt the series of facts, and often render it extremely difficult to reconcile them with each other: It has not, like the elements of geometry, the advantage of being a complete science, the parts of which are all closely connected together: Its actual progress, however, is so rapid, and the facts, under the modern doctrine, have assumed so happy an arrangement, that we have ground to hope, even in our own times, to see it approach near to the highest state of perfection of which it is susceptible.

The rigorous law from which I have never deviated, of forming no conclusions which are not fully warranted by experiment, and of never[Pg xxi] supplying the absence of facts, has prevented me from comprehending in this work the branch of chemistry which treats of affinities, although it is perhaps the best calculated of any part of chemistry for being reduced into a completely systematic body. Messrs Geoffroy, Gellert, Bergman, Scheele, De Morveau, Kirwan, and many others, have collected a number of particular facts upon this subject, which only wait for a proper arrangement; but the principal data are still wanting, or, at least, those we have are either not sufficiently defined, or not sufficiently proved, to become the foundation upon which to build so very important a branch of chemistry. This science of affinities, or elective attractions, holds the same place with regard to the other branches of chemistry, as the higher or transcendental geometry does with respect to the simpler and elementary part; and I thought it improper to involve those simple and plain elements, which I flatter myself the greatest part of my readers will easily understand, in the obscurities and difficulties which still attend that other very useful and necessary branch of chemical science.

Perhaps a sentiment of self-love may, without my perceiving it, have given additional force to[Pg xxii] these reflections. Mr de Morveau is at present engaged in publishing the article *Affinity* in the Methodical Encyclopædia; and I had more reasons than one to decline entering upon a work in which he is employed.

It will, no doubt, be a matter of surprise, that in a treatise upon the elements of chemistry, there should be no chapter on the constituent and elementary parts of matter; but I shall take occasion, in this place, to remark, that the fondness for reducing all the bodies in nature to three or four elements, proceeds from a prejudice which has descended to us from the Greek Philosophers. The notion of four elements, which, by the variety of their proportions, compose all the known substances in nature, is a mere hypothesis, assumed long before the first principles of experimental philosophy or of chemistry had any existence. In those days, without possessing facts, they framed systems; while we, who have collected facts, seem determined to reject them, when they do not agree with our prejudices. The authority of these fathers of human philosophy still carry great weight, and there is reason to fear that it will even bear hard upon generations yet to come.[Pg xxiii]

It is very remarkable, that, notwithstanding of the number of philosophical chemists who have supported the doctrine of the four elements, there is not one who has not been led by the evidence of facts to admit a greater number of elements into their theory. The first chemists that wrote after the revival of letters, considered sulphur and salt as elementary substances entering into the composition of a great number of substances; hence, instead of four, they admitted the existence of six elements. Beccher assumes the existence of three kinds of earth, from the combination of which, in different proportions, he supposed all the varieties of metallic substances to be produced. Stahl gave a new modification to this system; and succeeding chemists have taken the liberty to make or to imagine changes and additions of a similar nature. All these chemists were carried along by the influence of the genius of the age in which they lived, which contented itself with assertions without proofs; or, at least, often admitted as proofs the slighted degrees of probability, unsupported by that strictly rigorous analysis required by modern philosophy.[Pg xxiv]

All that can be said upon the number and nature of elements is, in my opinion, confined to discussions entirely of a metaphysical nature. The subject only furnishes us with indefinite problems, which may be solved in a thousand different ways, not one of which, in all probability, is consistent with nature. I shall therefore only add upon this subject, that if, by the term *elements*, we mean to express those simple and indivisible atoms of which matter is composed, it is extremely probable we know nothing at all about them; but, if we apply the term *elements*, or *principles of bodies*, to express our idea of the last point which analysis is capable of reaching, we must admit, as elements, all the substances into which we are capable, by any means, to reduce bodies by decomposition. Not that we are entitled to affirm, that these substances we consider as simple may not be compounded of two, or even of a greater number of principles; but, since these principles cannot be separated, or rather since we have not hitherto discovered the means of separating them, they act with regard to us as simple substances, and we ought never to suppose them compounded until experiment and observation has proved them to be so.[Pg xxv]

The foregoing reflections upon the progress of chemical ideas naturally apply to the words by which these ideas are to be expressed. Guided by the work which, in the year 1787, Messrs de Morveau, Berthollet, de Fourcroy, and I composed upon the Nomenclature of Chemistry, I have endeavoured, as much as possible, to denominate simple bodies by simple terms, and I was naturally led to name these first. It will be recollected, that we were obliged to retain that name of any substance by which it had been long known in the world, and that in two cases only we took the liberty of making alterations; first, in the case of those which were but newly discovered, and had not yet obtained names, or at least which had been known but for a short time, and the names of which had not yet received the sanction of the public; and, secondly, when the names which had been adopted, whether by the ancients or the moderns, appeared to us to express evidently false ideas, when they confounded the substances, to which they were applied, with others possessed of different, or perhaps opposite qualities. We made no scruple, in this case, of substituting other names in their room, and the greatest number of these were borrowed from the Greek language. We[Pg xxvi]endeavoured to frame them in such a manner as to express the most general and the most characteristic quality of the substances; and this was attended with the additional advantage both of assisting the memory of beginners, who find it difficult to remember a new word which has no meaning, and of accustoming them early to admit no word without connecting with it some determinate idea.

To those bodies which are formed by the union of several simple substances we gave new names, compounded in such a manner as the nature of the substances directed; but, as the number of double combinations is already very considerable, the only method by which we could avoid confusion, was to divide them into classes. In the natural order of ideas, the name of the class or genus is that which expresses a quality common to a great number of individuals: The name of the species, on the contrary, expresses a quality peculiar to certain individuals only.

These distinctions are not, as some may imagine, merely metaphysical, but are established by Nature. "A child," says the Abbé de Condillac,[Pg xxvii] "is taught to give the name *tree* to the first one which is pointed out to him. The next one he sees presents the same idea, and he gives it the same name. This he does likewise to a third and a fourth, till at last the word *tree*, which he first applied to an individual, comes to be employed by him as the name of a class or a genus, an abstract idea, which comprehends all trees in general. But, when he learns that all trees serve not the same purpose, that they do not all produce the same kind of fruit, he will soon learn to distinguish them by specific and particular names." This is the logic of all the sciences, and is naturally applied to chemistry.

The acids, for example, are compounded of two substances, of the order of those which we consider as simple; the one constitutes acidity, and is common to all acids, and, from this substance, the name of the class or the genus ought to be taken; the other is peculiar to each acid, and distinguishes it from the rest, and from this substance is to be taken the name of the species. But, in the greatest number of acids, the two constituent elements, the acidifying principle,[Pg xxviii] and that which it acidifies, may exist in different proportions, constituting all the possible points of equilibrium or of saturation. This is the case in the sulphuric and the sulphurous acids; and these two states of the same acid we have marked by varying the termination of the specific name.

Metallic substances which have been exposed to the joint action of the air and of fire, lose their metallic lustre, increase in weight, and assume an earthy appearance. In this state, like the acids, they are compounded of a principle which is common to all, and one which is peculiar to each. In the same way, therefore, we have thought proper to class them under a generic name, derived from the common principle; for which purpose, we adopted the term *oxyd*; and we distinguish them from each other by the particular name of the metal to which each belongs.

Combustible substances, which in acids and metallic oxyds are a specific and particular principle, are capable of becoming, in their turn, common principles of a great number of substances. The sulphurous combinations have[Pg xxix] been long the only known ones in this kind. Now, however, we know, from the experiments of Messrs Vandermonde, Monge, and Berthollet, that charcoal may be combined with iron, and perhaps with several other metals; and that, from this combination, according to the proportions, may be produced steel, plumbago, &c. We know likewise, from the experiments of M. Pelletier, that phosphorus may be combined with a great number of metallic substances. These different combinations we have classed under generic names taken from the common substance, with a termination which marks this analogy, specifying them by another name taken from that substance which is proper to each.

The nomenclature of bodies compounded of three simple substances was attended with still greater difficulty, not only on account of their number, but, particularly, because we cannot express the nature of their constituent principles without employing more compound names. In the bodies which form this class, such as the neutral salts, for instance, we had to consider, 1st, The acidifying principle, which is common to them all; 2d, The acidifiable principle which constitutes their peculiar acid; 3d, The saline,[Pg xxx] earthy, or metallic basis, which determines the particular species of salt. Here we derived the name of each class of salts from the name of the acidifiable principle common to all the individuals of that class; and distinguished each species by the name of the saline, earthy, or metallic basis, which is peculiar to it.

A salt, though compounded of the same three principles, may, nevertheless, by the mere difference of their proportion, be in three different states. The nomenclature we have adopted would have been defective, had it not expressed these different states; and this we attained chiefly by changes of termination uniformly applied to the same state of the different salts.

In short, we have advanced so far, that from the name alone may be instantly found what the combustible substance is which enters into any combination; whether that combustible substance be combined with the acidifying principle, and in what proportion; what is the state of the acid; with what basis it is united; whether the saturation be exact, or whether the acid or the basis be in excess.[Pg xxxi]

It may be easily supposed that it was not possible to attain all these different objects without departing, in some instances, from established custom, and adopting terms which at first sight will appear uncouth and barbarous. But we considered that the ear is soon habituated to new words, especially when they are connected with a general and rational system. The names, besides, which were formerly employed, such as *powder of algaroth, salt of alembroth, pompholix, phagadenic water, turbith mineral,colcathar*, and many others, were neither less barbarous nor less uncommon. It required a great deal of practice, and no small degree of memory, to recollect the substances to which they were applied, much more to recollect the genus of combination to which they belonged. The names of *oil of tartar per deliquium, oil of vitriol, butter of arsenic and of antimony, flowers of zinc*, &c. were still more improper, because they suggested false ideas: For, in the whole mineral kingdom, and particularly in the metallic class, there exists no such thing as butters, oils, or flowers; and, in short, the substances to which they give these fallacious names, are nothing less than rank poisons.[Pg xxxii]

When we published our essay on the nomenclature of chemistry, we were reproached for having changed the language which was spoken by our masters, which they distinguished by their authority, and handed down to us. But those who reproach us on this account, have forgotten that it was Bergman and Macquer themselves who urged us to make this reformation. In a letter which the learned Professor of Upsal, M. Bergman, wrote, a short time before he died, to M. de Morveau, he bids him *spare no improper names; those who are learned, will always be learned, and those who are ignorant will thus learn sooner.*

There is an objection to the work which I am going to present to the public, which is perhaps better founded, that I have given no account of the opinion of those who have gone before me; that I have stated only my own opinion, without examining that of others. By this I have been prevented from doing that justice to my associates, and more especially to foreign chemists, which I wished to render them. But I beseech the reader to consider, that, if I had filled an elementary work with a multitude of quotations; if I had allowed myself to enter into[Pg xxxiii] long dissertations on the history of the science, and the works of those who have studied it, I must have lost sight of the true object I had in view, and produced a work, the reading of which must have been extremely tiresome to beginners. It is not to the history of the science, or of the human mind, that we are to attend in an elementary treatise: Our only aim ought to be ease and perspicuity, and with the utmost care to keep every thing out of view which might draw aside the attention of the student; it is a road which we should be continually rendering more smooth, and from which we should endeavour to remove every obstacle which can occasion delay. The sciences, from their own nature, present a sufficient number of difficulties, though we add not those which are foreign to them. But, besides this, chemists will easily perceive, that, in the first part of my work, I make very little use of any experiments but those which were made by myself: If at any time I have adopted, without acknowledgment, the experiments or the opinions of M. Berthollet, M. Fourcroy, M. de la Place, M. Monge, or, in general, of any of those whose principles are the same with my own, it is owing to this circumstance, that frequent intercourse, and the habit of communicating our[Pg xxxiv] ideas, our observations, and our way of thinking to each other, has established between us a sort of community of opinions, in which it is often difficult for every one to know his own.

The remarks I have made on the order which I thought myself obliged to follow in the arrangement of proofs and ideas, are to be applied only to the first part of this work. It is the only one which contains the general sum of the doctrine I have adopted, and to which I wished to give a form completely elementary.

The second part is composed chiefly of tables of the nomenclature of the neutral salts. To these I have only added general explanations, the object of which was to point out the most simple processes for obtaining the different kinds of known acids. This part contains nothing which I can call my own, and presents only a very short abridgment of the results of these processes, extracted from the works of different authors.

In the third part, I have given a description, in detail, of all the operations connected with modern chemistry. I have long thought that a[Pg xxxv] work of this kind was much wanted, and I am convinced it will not be without use. The method of performing experiments, and particularly those of modern chemistry, is not so generally known as it ought to be; and had I, in the different memoirs which I have presented to the Academy, been more particular in the detail of the manipulations of my experiments, it is probable I should have made myself better understood, and the science might have made a more rapid progress. The order of the different matters contained in this third part appeared to me to be almost arbitrary; and the only one I have observed was to class together, in each of the chapters of which it is composed, those operations which are most connected with one another. I need hardly mention that this part could not be borrowed from any other work, and that, in the principal articles it contains, I could not derive assistance from any thing but the experiments which I have made myself.

I shall conclude this preface by transcribing, literally, some observations of the Abbé de Condillac, which I think describe, with a good deal of truth, the state of chemistry at a period not far distant from our own. These observations[Pg xxxvi] were made on a different subject; but they will not, on this account, have less force, if the application of them be thought just.

'Instead of applying observation to the things we wished to know, we have chosen rather to imagine them. Advancing from one ill founded supposition to another, we have at last bewildered ourselves amidst a multitude of errors. These errors becoming prejudices, are, of course, adopted as principles, and we thus bewilder ourselves more and more. The method, too, by which we conduct our reasonings is as absurd; we abuse words which we do not understand, and call this the art of reasoning. When matters have been brought this length, when errors have been thus accumulated, there is but one remedy by which order can be restored to the faculty of thinking; this is, to forget all that we have learned, to trace back our ideas to their source, to follow the train in which they rise, and, as my Lord Bacon says, to frame the human understanding anew.

'This remedy becomes the more difficult in proportion as we think ourselves more learned.[Pg xxxvii] Might it not be thought that works which treated of the sciences with the utmost perspicuity, with great precision and order, must be understood by every body? The fact is, those who have never studied any thing will understand them better than those who have studied a great deal, and especially than those who have written a great deal.'

At the end of the fifth chapter, the Abbé de Condillac adds: 'But, after all, the sciences have made progress, because philosophers have applied themselves with more attention to observe, and have communicated to their language that precision and accuracy which they have employed in their observations: In correcting their language they reason better.'

ELEMENTS
OF
CHEMISTRY.

PART I.

Of the Formation and Decomposition of Aëriform Fluids—of the Combustion of Simple Bodies—and the Formation of Acids.

CHAP. I.
Of the Combinations of Caloric, and the Formation of Elastic Aëriform Fluids.

That every body, whether solid or fluid, is augmented in all its dimensions by any increase of its sensible heat, was long ago fully established as a physical axiom, or universal proposition, by the celebrated Boerhaave. Such facts as have been adduced for controverting the[Pg 2] generality of this principle offer only fallacious results, or, at least, such as are so complicated with foreign circumstances as to mislead the judgment: But, when we separately consider the effects, so as to deduce each from the cause to which they separately belong, it is easy to perceive that the separation of particles by heat is a constant and general law of nature.

When we have heated a solid body to a certain degree, and have thereby caused its particles to separate from each other, if we allow the body to cool, its particles again approach each other in the same proportion in which they were separated by the increased temperature; the body returns through the same degrees of expansion which it before extended through; and, if it be brought back to the same temperature from which we set out at the commencement of the experiment, it recovers exactly the same dimensions which it formerly occupied. But, as we are still very far from being able to arrive at the degree of absolute cold, or deprivation of all heat, being unacquainted with any degree of coldness which we cannot suppose capable of still farther augmentation, it follows, that we are still incapable of causing the ultimate particles of bodies to approach each other as near as is possible; and, consequently, that the particles of all bodies do not touch each other in any state hitherto known, which, tho'[Pg 3] a very singular conclusion, is yet impossible to be denied.

It is supposed, that, since the particles of bodies are thus continually impelled by heat to separate from each other, they would have no connection between themselves; and, of consequence, that there could be no solidity in nature, unless they were held together by some other power which tends to unite them, and, so to speak, to chain them together; which power, whatever be its cause, or manner of operation, we name Attraction.

Thus the particles of all bodies may be considered as subjected to the action of two opposite powers, the one repulsive, the other attractive, between which they remain in equilibrio. So long as the attractive force remains stronger, the body must continue in a state of solidity; but if, on the contrary, heat has so far removed these particles from each other, as to place them beyond the sphere of attraction, they lose the adhesion they before had with each other, and the body ceases to be solid.

Water gives us a regular and constant example of these facts; whilst below Zero[2] of the French thermometer, or 32° of Fahrenheit,[Pg 4] it remains solid, and is called ice. Above that degree of temperature, its particles being no longer held together by reciprocal attraction, it becomes liquid; and, when we raise its temperature above 80°, (212°) its particles, giving way to the repulsion caused by the heat, assume the state of vapour or gas, and the water is changed into an aëriform fluid.

The same may be affirmed of all bodies in nature: They are either solid or liquid, or in the state of elastic aëriform vapour, according to the proportion which takes place between the attractive force inherent in their particles, and the repulsive power of the heat acting upon these; or, what amounts to the same thing, in proportion to the degree of heat to which they are exposed.

It is difficult to comprehend these phenomena, without admitting them as the effects of a real and material substance, or very subtile fluid, which, insinuating itself between the particles of bodies, separates them from each other; and, even allowing the existence of this fluid to be hypothetical, we shall see in the sequel, that it explains the phenomena of nature in a very satisfactory manner.

This substance, whatever it is, being the cause of heat, or, in other words, the sensation which we call *warmth* being caused by the accumulation of this substance, we cannot, in strict language,[Pg 5] distinguish it by the term *heat*; because the same name would then very improperly express both cause and effect. For this reason, in the memoir which I published in 1777[3], I gave it the names of *igneous fluid* and *matter of heat*. And, since that time, in the work[4] published by Mr de Morveau, Mr Berthollet, Mr de Fourcroy, and myself, upon the reformation of chemical nomenclature, we thought it necessary to banish all periphrastic expressions, which both lengthen physical language, and render it more tedious and less distinct, and which even frequently does not convey sufficiently just ideas of the subject intended. Wherefore, we have distinguished the cause of heat, or that exquisitely elastic fluid which produces it, by the term of *caloric*. Besides, that this expression fulfils our object in the system which we have adopted, it possesses this farther advantage, that it accords with every species of opinion, since, strictly speaking, we are not obliged to suppose this to be a real substance; it being sufficient, as will more clearly appear in the sequel of this work, that it be considered as the repulsive cause, whatever that may be, which separates the particles of matter from each other; so that[Pg 6] we are still at liberty to investigate its effects in an abstract and mathematical manner.

In the present state of our knowledge, we are unable to determine whether light be a modification of caloric, or if caloric be, on the contrary, a modification of light. This, however, is indisputable, that, in a system where only decided facts are admissible, and where we avoid, as far as possible, to suppose any thing to be that is not really known to exist, we ought provisionally to distinguish, by distinct terms, such things as are known to produce different effects. We therefore distinguish light from caloric; though we do not therefore deny that these have certain qualities in common, and that, in certain circumstances, they combine with other bodies almost in the same manner, and produce, in part, the same effects.

What I have already said may suffice to determine the idea affixed to the word *caloric*; but there remains a more difficult attempt, which is, to give a just conception of the manner in which caloric acts upon other bodies. Since this subtile matter penetrates through the pores of all known substances; since there are no vessels through which it cannot escape, and, consequently, as there are none which are capable of retaining it, we can only come at the knowledge of its properties by effects which are fleeting, and difficultly ascertainable. It is in[Pg 7] these things which we neither see nor feel, that it is especially necessary to guard against the extravagancy of our imagination, which forever inclines to step beyond the bounds of truth, and is very difficultly restrained within the narrow line of facts.

We have already seen, that the same body becomes solid, or fluid, or aëriform, according to the quantity of caloric by which it is penetrated; or, to speak more strictly, according as the repulsive force exerted by the caloric is equal to, stronger, or weaker, than the attraction of the particles of the body it acts upon.

But, if these two powers only existed, bodies would become liquid at an indivisible degree of the thermometer, and would almost instantaneously pass from the solid state of aggregation to that of aëriform elasticity. Thus water, for instance, at the very moment when it ceases to be ice, would begin to boil, and would be transformed into an aëriform fluid, having its particles scattered indefinitely through the surrounding space. That this does not happen, must depend upon the action of some third power. The pressure of the atmosphere prevents this separation, and causes the water to remain in the liquid state till it be raised to $80°$ of temperature ($212°$) above zero of the French thermometer, the quantity of caloric which it receives in the lowest temperature being insufficient[Pg 8] to overcome the pressure of the atmosphere.

Whence it appears that, without this atmospheric pressure, we should not have any permanent liquid, and should only be able to see bodies in that state of existence in the very instant of melting, as the smallest additional caloric would instantly separate their particles, and dissipate them through the surrounding medium. Besides, without this atmospheric pressure, we should not even have any aëriform fluids, strictly speaking, because the moment the force of attraction is overcome by the repulsive power of the caloric, the particles would separate themselves indefinitely, having nothing to give limits to their expansion, unless their own gravity might collect them together, so as to form an atmosphere.

Simple reflection upon the most common experiments is sufficient to evince the truth of these positions. They are more particularly proved by the following experiment, which I published in the Memoirs of the French Academy for 1777, p. 426.

Having filled with sulphuric ether[5] a small narrow glass vessel, A, (Plate VII. Fig. 17.), standing[Pg 9] upon its stalk P, the vessel, which is from twelve to fifteen lines diameter, is to be covered by a wet bladder, tied round its neck with several turns of strong thread; for greater security, fix a second bladder over the first. The vessel should be filled in such a manner with the ether, as not to leave the smallest portion of air between the liquor and the bladder. It is now to be placed under the recipient BCD of an air-pump, of which the upper part B ought to be fitted with a leathern lid, through which passes a wire EF, having its point F very sharp; and in the same receiver there ought to be placed the barometer GH. The whole being thus disposed, let the recipient be exhausted, and then, by pushing down the wire EF, we make a hole in the bladder. Immediately the ether begins to boil with great violence, and is changed into an elastic aëriform fluid, which fills the receiver. If the quantity of ether be sufficient to leave a few drops in the phial after the evaporation is finished, the elastic fluid produced will sustain the mercury in the barometer attached to the air-pump, at eight or ten inches in winter, and from[Pg 10] twenty to twenty-five in summer[6]. To render this experiment more complete, we may introduce a small thermometer into the phial A, containing the ether, which will descend considerably during the evaporation.

The only effect produced in this experiment is, the taking away the weight of the atmosphere, which, in its ordinary state, presses on the surface of the ether; and the effects resulting from this removal evidently prove, that, in the ordinary temperature of the earth, ether would always exist in an aëriform state, but for the pressure of the atmosphere, and that the passing of the ether from the liquid to the aëriform state is accompanied by a considerable lessening of heat; because, during the evaporation, a part of the caloric, which was before in a free state, or at least in equilibrio in the surrounding bodies, combines with the ether, and causes it to assume the aëriform state.

The same experiment succeeds with all evaporable fluids, such as alkohol, water, and even mercury; with this difference, that the atmosphere formed in the receiver by alkohol only[Pg 11] supports the attached barometer about one inch in winter, and about four or five inches in summer; that formed by water, in the same situation, raises the mercury only a few lines, and that by quicksilver but a few fractions of a line. There is therefore less fluid evaporated from alkohol than from ether, less from water than from alkohol, and still less from mercury than from either; consequently there is less caloric employed, and less cold produced, which quadrates exactly with the results of these experiments.

Another species of experiment proves very evidently that the aëriform state is a modification of bodies dependent on the degree of temperature, and on the pressure which these bodies undergo. In a Memoir read by Mr de la Place and me to the Academy in 1777, which has not been printed, we have shown, that, when ether is subjected to a pressure equal to twenty-eight inches of the barometer, or about the medium pressure of the atmosphere, it boils at the temperature of about 32° (104°), or 33° (106.25°), of the thermometer. Mr de Luc, who has made similar experiments with spirit of wine, finds it boils at 67° (182.75°). And all the world knows that water boils at 80° (212°). Now, boiling being only the evaporation of a liquid, or the moment of its passing from the fluid to the aëriform state, it is evident that, if we keep[Pg 12]ether continually at the temperature of 33° (106.25°), and under the common pressure of the atmosphere, we shall have it always in an elastic aëriform state; and that the same thing will happen with alkohol when above 67° (182.75°), and with water when above 80° (212°); all which are perfectly conformable to the following experiment[7].

I filled a large vessel ABCD (Plate VII. Fig. 16.) with water, at 35° (110.75°), or 36° (113°); I suppose the vessel transparent, that we may see what takes place in the experiment; and we can easily hold the hands in water at that temperature without inconvenience. Into it I plunged some narrow necked bottles F, G, which were filled with the water, after which they were turned up, so as to rest on their mouths on the bottom of the vessel. Having next put some ether into a very small matrass, with its neck *a b c*, twice bent as in the Plate, I plunged this matrass into the water, so as to have its neck inserted into the mouth of one of the bottles F. Immediately upon feeling the effects of the heat communicated to it by the water in the vessel ABCD it began to boil; and the caloric entering into combination with it, changed it into elastic aëriform fluid, with which I filled several bottles successively, F, G, &c.

[Pg 13]
This is not the place to enter upon the examination of the nature and properties of this aëriform fluid, which is extremely inflammable; but, confining myself to the object at present in view, without anticipating circumstances, which I am not to suppose the reader to know, I shall only observe, that the ether, from this experiment, is almost only capable of existing in the aëriform state in our world; for, if the weight of our atmosphere was only equal to between 20 and 24 inches of the barometer, instead of 28 inches, we should never be able to obtain ether in the liquid state, at least in summer; and the formation of ether would consequently be impossible upon mountains of a moderate degree of elevation, as it would be converted into gas immediately upon being produced, unless we employed recipients of extraordinary strength, together with refrigeration and compression. And, lastly, the temperature of the blood being nearly that at which ether passes from the liquid to the aëriform state, it must evaporate in the primae viae, and consequently it is very probable the medical properties of this fluid depend chiefly upon its mechanical effect.

These experiments succeed better with nitrous ether, because it evaporates in a lower temperature than sulphuric ether. It is more difficult to obtain alkohol in the aëriform state; because, as it requires 67° (182.75°) to reduce it to vapour,[Pg 14] the water of the bath must be almost boiling, and consequently it is impossible to plunge the hands into it at that temperature.

It is evident that, if water were used in the foregoing experiment, it would be changed into gas, when exposed to a temperature superior to that at which it boils. Although thoroughly convinced of this, Mr de la Place and myself judged it necessary to confirm it by the following direct experiment. We filled a glass jar A, (Plate VII. Fig. 5.) with mercury, and placed it with its mouth downwards in a dish B, likewise filled with mercury, and having introduced about two gross of water into the jar, which rose to the top of the mercury at CD; we then plunged the whole apparatus into an iron boiler EFGH, full of boiling sea-water of the temperature of 85° (123.25°), placed upon the furnace GHIK. Immediately upon the water over the mercury attaining the temperature of 80° (212°), it began to boil; and, instead of only filling the small space ACD, it was converted into an aëriform fluid, which filled the whole jar; the mercury even descended below the surface of that in the dish B; and the jar must have been overturned, if it had not been very thick and heavy, and fixed to the dish by means of iron-wire. Immediately after withdrawing the apparatus from the boiler, the vapour in the jar began to condense, and the[Pg 15] mercury rose to its former station; but it returned again to the aëriform state a few seconds after replacing the apparatus in the boiler.

We have thus a certain number of substances, which are convertible into elastic aëriform fluids by degrees of temperature, not much superior to that of our atmosphere. We shall afterwards find that there are several others which undergo the same change in similar circumstances, such as muriatic or marine acid, ammoniac or volatile alkali, the carbonic acid or fixed air, the sulphurous acid, &c. All of these are permanently elastic in or about the mean temperature of the atmosphere, and under its common pressure.

All these facts, which could be easily multiplied if necessary, give me full right to assume, as a general principle, that almost every body in nature is susceptible of three several states of existence, solid, liquid, and aëriform, and that these three states of existence depend upon the quantity of caloric combined with the body. Henceforwards I shall express these elastic aëriform fluids by the generic term *gas*; and in each species of gas I shall distinguish between the caloric, which in some measure serves the purpose of a solvent, and the substance, which in combination with the caloric, forms the base of the gas.[Pg 16]

To these bases of the different gases, which are hitherto but little known, we have been obliged to assign names; these I shall point out in Chap. IV. of this work, when I have previously given an account of the phenomena attendant upon the heating and cooling of bodies, and when I have established precise ideas concerning the composition of our atmosphere.

We have already shown, that the particles of every substance in nature exist in a certain state of equilibrium, between that attraction which tends to unite and keep the particles together, and the effects of the caloric which tends to separate them. Hence the caloric not only surrounds the particles of all bodies on every side, but fills up every interval which the particles of bodies leave between each other. We may form an idea of this, by supposing a vessel filled with small spherical leaden bullets, into which a quantity of fine sand is poured, which, insinuating into the intervals between the bullets, will fill up every void. The balls, in this comparison, are to the sand which surrounds them exactly in the same situation as the particles of bodies are with respect to the caloric; with this difference only, that the balls are supposed to touch each other, whereas the particles of bodies are not in contact, being retained at a small distance from each other, by the caloric.[Pg 17]

If, instead of spherical balls, we substitute solid bodies of a hexahedral, octohedral, or any other regular figure, the capacity of the intervals between them will be lessened, and consequently will no longer contain the same quantity of sand. The same thing takes place, with respect to natural bodies; the intervals left between their particles are not of equal capacity, but vary in consequence of the different figures and magnitude of their particles, and of the distance at which these particles are maintained, according to the existing proportion between their inherent attraction, and the repulsive force exerted upon them by the caloric.

In this manner we must understand the following expression, introduced by the English philosophers, who have given us the first precise ideas upon this subject; *the capacity of bodies for containing the matter of heat*. As comparisons with sensible objects are of great use in assisting us to form distinct notions of abstract ideas, we shall endeavour to illustrate this, by instancing the phenomena which take place between water and bodies which are wetted and penetrated by it, with a few reflections.

If we immerge equal pieces of different kinds of wood, suppose cubes of one foot each, into water, the fluid gradually insinuates itself into their pores, and the pieces of wood are augmented both in weight and magnitude: But[Pg 18] each species of wood will imbibe a different quantity of water; the lighter and more porous woods will admit a larger, the compact and closer grained will admit a lesser quantity; for the proportional quantities of water imbibed by the pieces will depend upon the nature of the constituent particles of the wood, and upon the greater or lesser affinity subsisting between them and water. Very resinous wood, for instance, though it may be at the same time very porous, will admit but little water. We may therefore say, that the different kinds of wood possess different capacities for receiving water; we may even determine, by means of the augmentation of their weights, what quantity of water they have actually absorbed; but, as we are ignorant how much water they contained, previous to immersion, we cannot determine the absolute quantity they contain, after being taken out of the water.

The same circumstances undoubtedly take place, with bodies that are immersed in caloric; taking into consideration, however, that water is an incompressible fluid, whereas caloric is, on the contrary, endowed with very great elasticity; or, in other words, the particles of caloric have a great tendency to separate from each other, when forced by any other power to approach; this difference must of necessity occasion[Pg 19]very considerable diversities in the results of experiments made upon these two substances.

Having established these clear and simple propositions, it will be very easy to explain the ideas which ought to be affixed to the following expressions, which are by no means synonymous, but possess each a strict and determinate meaning, as in the following definitions:

Free caloric, is that which is not combined in any manner with any other body. But, as we live in a system to which caloric has a very strong adhesion, it follows that we are never able to obtain it in the state of absolute freedom.

Combined caloric, is that which is fixed in bodies by affinity or elective attraction, so as to form part of the substance of the body, even part of its solidity.

By the expression *specific caloric* of bodies, we understand the respective quantities of caloric requisite for raising a number of bodies of the same weight to an equal degree of temperature. This proportional quantity of caloric depends upon the distance between the constituent particles of bodies, and their greater or lesser degrees of cohesion; and this distance, or rather the space or void resulting from it, is, as I have already observed, called the *capacity of bodies for containing caloric.*[Pg 20]

Heat, considered as a sensation, or, in other words, sensible heat, is only the effect produced upon our sentient organs, by the motion or passage of caloric, disengaged from the surrounding bodies. In general, we receive impressions only in consequence of motion, and we might establish it as an axiom, *That*, WITHOUT MOTION, THERE IS NO SENSATION. This general principle applies very accurately to the sensations of heat and cold: When we touch a cold body, the caloric which always tends to become in equilibrio in all bodies, passes from our hand into the body we touch, which gives us the feeling or sensation of cold. The direct contrary happens, when we touch a warm body, the caloric then passing from the body into our hand, produces the sensation of heat. If the hand and the body touched be of the same temperature, or very nearly so, we receive no impression, either of heat or cold, because there is no motion or passage of caloric; and thus no sensation can take place, without some correspondent motion to occasion it.

When the thermometer rises, it shows, that free caloric is entering into the surrounding bodies: The thermometer, which is one of these, receives its share in proportion to its mass, and to the capacity which it possesses for containing caloric. The change therefore which takes place upon the thermometer, only announces a[Pg 21]change of place of the caloric in those bodies, of which the thermometer forms one part; it only indicates the portion of caloric received, without being a measure of the whole quantity disengaged, displaced, or absorbed.

The most simple and most exact method for determining this latter point, is that described by Mr de la Place, in the Memoirs of the Academy, No. 1780, p. 364; a summary explanation of which will be found towards the conclusion of this work. This method consists in placing a body, or a combination of bodies, from which caloric is disengaging, in the midst of a hollow sphere of ice; and the quantity of ice melted becomes an exact measure of the quantity of caloric disengaged. It is possible, by means of the apparatus which we have caused to be constructed upon this plan, to determine, not as has been pretended, the capacity of bodies for containing heat, but the ratio of the increase or diminution of capacity produced by determinate degrees of temperature. It is easy with the same apparatus, by means of divers combinations of experiments, to determine the quantity of caloric requisite for converting solid substances into liquids, and liquids into elastic aëriform fluids; and, *vice versa*, what quantity of caloric escapes from elastic vapours in changing to liquids, and what quantity escapes from liquids during their conversion into solids. Perhaps,[Pg 22] when experiments have been made with sufficient accuracy, we may one day be able to determine the proportional quantity of caloric, necessary for producing the several species of gasses. I shall hereafter, in a separate chapter, give an account of the principal results of such experiments as have been made upon this head.

It remains, before finishing this article, to say a few words relative to the cause of the elasticity of gasses, and of fluids in the state of vapour. It is by no means difficult to perceive that this elasticity depends upon that of caloric, which seems to be the most eminently elastic body in nature. Nothing is more readily conceived, than that one body should become elastic by entering into combination with another body possessed of that quality. We must allow that this is only an explanation of elasticity, by an assumption of elasticity, and that we thus only remove the difficulty one step farther, and that the nature of elasticity, and the reason for caloric being elastic, remains still unexplained. Elasticity in the abstract is nothing more than that quality of the particles of bodies by which they recede from each other when forced together. This tendency in the particles of caloric to separate, takes place even at considerable distances. We shall be satisfied of this, when we consider that air is susceptible of undergoing great compression, which supposes that its particles[Pg 23] were previously very distant from each other; for the power of approaching together certainly supposes a previous distance, at least equal to the degree of approach. Consequently, those particles of the air, which are already considerably distant from each other, tend to separate still farther. In fact, if we produce Boyle's vacuum in a large receiver, the very last portion of air which remains spreads itself uniformly through the whole capacity of the vessel, however large, fills it completely throughout, and presses every where against its sides: We cannot, however, explain this effect, without supposing that the particles make an effort to separate themselves on every side, and we are quite ignorant at what distance, or what degree of rarefaction, this effort ceases to act.

Here, therefore, exists a true repulsion between the particles of elastic fluids; at least, circumstances take place exactly as if such a repulsion actually existed; and we have very good right to conclude, that the particles of caloric mutually repel each other. When we are once permitted to suppose this repelling force, the *rationale* of the formation of gasses, or aëriform fluids, becomes perfectly simple; tho' we must, at the same time, allow, that it is extremely difficult to form an accurate conception of this repulsive force acting upon very minute[Pg 24] particles placed at great distances from each other.

It is, perhaps, more natural to suppose, that the particles of caloric have a stronger mutual attraction than those of any other substance, and that these latter particles are forced asunder in consequence of this superior attraction between the particles of the caloric, which forces them between the particles of other bodies, that they may be able to reunite with each other. We have somewhat analogous to this idea in the phenomena which occur when a dry sponge is dipt into water: The sponge swells; its particles separate from each other; and all its intervals are filled up by the water. It is evident, that the sponge, in the act of swelling, has acquired a greater capacity for containing water than it had when dry. But we cannot certainly maintain, that the introduction of water between the particles of the sponge has endowed them with a repulsive power, which tends to separate them from each other; on the contrary, the whole phenomena are produced by means of attractive powers; and these are, *first*, The gravity of the water, and the power which it exerts on every side, in common with all other fluids; *2dly*, The force of attraction which takes place between the particles of the water, causing them to unite together; *3dly*, The mutual attraction of the particles of the sponge with each other;[Pg 25] and, *lastly*, The reciprocal attraction which exists between the particles of the sponge and those of the water. It is easy to understand, that the explanation of this fact depends upon properly appreciating the intensity of, and connection between, these several powers. It is probable, that the separation of the particles of bodies, occasioned by caloric, depends in a similar manner upon a certain combination of different attractive powers, which, in conformity with the imperfection of our knowledge, we endeavour to express by saying, that caloric communicates a power of repulsion to the particles of bodies.

FOOTNOTES:

[2]Whenever the degree of heat occurs in this work, it is stated by the author according to Reaumur's scale. The degrees within brackets are the correspondent degrees of Fahrenheit's scale, added by the translator. E.

[3]Collections of the French Academy of Sciences for that year, p. 420.

[4]Chemical Nomenclature.

[5]As I shall afterwards give a definition, and explain the properties of the liquor called *ether*, I shall only premise here, that it is a very volatile inflammable liquor, having a considerably smaller specific gravity than water, or even spirit of wine.—A.

[6]It would have been more satisfactory if the Author had specified the degrees of the thermometer at which these heights of the mercury in the barometer are produced.

[7]Vide Memoirs of the French Academy, anno 1780, p. 335.—A.

[Pg 26]

CHAP. II.

General Views relative to the Formation and Composition of our Atmosphere.

These views which I have taken of the formation of elastic aëriform fluids or gasses, throw great light upon the original formation of the atmospheres of the planets, and particularly that of our earth. We readily conceive, that it must necessarily consist of a mixture of the following substances: *First*, Of all bodies that are susceptible of evaporation, or, more strictly speaking, which are capable of retaining the state of aëriform elasticity in the temperature of our atmosphere, and under a pressure equal to that of a column of twenty-eight inches of quicksilver in the barometer; and, *secondly*, Of all substances, whether liquid or solid, which are capable of being dissolved by this mixture of different gasses.

The better to determine our ideas relating to this subject, which has not hitherto been sufficiently considered, let us, for a moment, conceive what change would take place in the various[Pg 27] substances which compose our earth, if its temperature were suddenly altered. If, for instance, we were suddenly transported into the region of the planet Mercury, where probably the common temperature is much superior to that of boiling water, the water of the earth, and all the other fluids which are susceptible of the gasseous state, at a temperature near to that of boiling water, even quicksilver itself, would become rarified; and all these substances would be changed into permanent aëriform fluids or gasses, which would become part of the new atmosphere. These new species of airs or gasses would mix with those already existing, and certain reciprocal decompositions and new combinations would take place, until such time as all the elective attractions or affinities subsisting amongst all these new and old gasseous substances had operated fully; after which, the elementary principles composing these gasses, being saturated, would remain at rest. We must attend to this, however, that, even in the above hypothetical situation, certain bounds would occur to the evaporation of these substances, produced by that very evaporation itself; for as, in proportion to the increase of elastic fluids, the pressure of the atmosphere would be augmented, as every degree of pressure tends, in some measure, to prevent evaporation, and as even the most evaporable[Pg 28] fluids can resist the operation of a very high temperature without evaporating, if prevented by a proportionally stronger compression, water and all other liquids being able to sustain a red heat in Papin's digester; we must admit, that the new atmosphere would at last arrive at such a degree of weight, that the water which had not hitherto evaporated would cease to boil, and, of consequence, would remain liquid; so that, even upon this supposition, as in all others of the same nature, the increasing gravity of the atmosphere would find certain limits which it could not exceed. We might even extend these reflections greatly farther, and examine what change might be produced in such situations upon stones, salts, and the greater part of the fusible substances which compose the mass of our earth. These would be softened, fused, and changed into fluids, &c.: But these speculations carry me from my object, to which I hasten to return.

By a contrary supposition to the one we have been forming, if the earth were suddenly transported into a very cold region, the water which at present composes our seas, rivers, and springs, and probably the greater number of the fluids we are acquainted with, would be converted into solid mountains and hard rocks, at first diaphanous[Pg 29] and homogeneous, like rock crystal, but which, in time, becoming mixed with foreign and heterogeneous substances, would become opake stones of various colours. In this case, the air, or at least some part of the aëriform fluids which now compose the mass of our atmosphere, would doubtless lose its elasticity for want of a sufficient temperature to retain them in that state: They would return to the liquid state of existence, and new liquids would be formed, of whose properties we cannot, at present, form the most distant idea.

These two opposite suppositions give a distinct proof of the following corollaries:*First*, That *solidity*, *liquidity*, and *aëriform elasticity*, are only three different states of existence of the same matter, or three particular modifications which almost all substances are susceptible of assuming successively, and which solely depend upon the degree of temperature to which they are exposed; or, in other words, upon the quantity of caloric with which they are penetrated[8]. *2dly*, That it is extremely probable that air is a fluid naturally existing in a state of vapour; or, as we may better express it, that our atmosphere is a compound of all the fluids[Pg 30] which are susceptible of the vaporous or permanently elastic state, in the usual temperature, and under the common pressure. *3dly*, That it is not impossible we may discover, in our atmosphere, certain substances naturally very compact, even metals themselves; as a metallic substance, for instance, only a little more volatile than mercury, might exist in that situation.

Amongst the fluids with which we are acquainted, some, as water and alkohol, are susceptible of mixing with each other in all proportions; whereas others, on the contrary, as quicksilver, water, and oil, can only form a momentary union; and, after being mixed together, separate and arrange themselves according to their specific gravities. The same thing ought to, or at least may, take place in the atmosphere. It is possible, and even extremely probable, that, both at the first creation, and every day, gasses are formed, which are difficultly miscible with atmospheric air, and are continually separating from it. If these gasses be specifically lighter than the general atmospheric mass, they must, of course, gather in the higher regions, and form strata that float upon the common air. The phenomena which accompany igneous meteors induce me to believe, that there exists in the upper parts[Pg 31] of our atmosphere a stratum of inflammable fluid in contact with those strata of air which produce the phenomena of the aurora borealis and other fiery meteors.—I mean hereafter to pursue this subject in a separate treatise.

FOOTNOTES:

[8]The degree of pressure which they undergo must be taken into account. E.

CHAP. III.

[Pg 32]
Analysis of Atmospheric Air, and its Division into two Elastic Fluids; the one fit for Respiration, the other incapable of being respired.

From what has been premised, it follows, that our atmosphere is composed of a mixture of every substance capable of retaining the gaseous or aëriform state in the common temperature, and under the usual pressure which it experiences. These fluids constitute a mass, in some measure homogeneous, extending from the surface of the earth to the greatest height hitherto attained, of which the density continually decreases in the inverse ratio of the superincumbent weight. But, as I have before observed, it is possible that this first stratum is surrounded by several others consisting of very different fluids.

Our business, in this place, is to endeavour to determine, by experiments, the nature of the elastic fluids which compose the inferior stratum of air which we inhabit. Modern chemistry has made great advances in this research; and it will appear by the following details that the analysis of atmospherical air has been more[Pg 33] rigorously determined than that of any other substance of the class. Chemistry affords two general methods of determining the constituent principles of bodies, the method of analysis, and that of synthesis. When, for instance, by combining water with alkohol, we form the species of liquor called, in commercial language, brandy or spirit of wine, we certainly have a right to conclude, that brandy, or spirit of wine, is composed of alkohol combined with water. We can produce the same result by the analytical method; and in general it ought to be considered as a principle in chemical science, never to rest satisfied without both these species of proofs.

We have this advantage in the analysis of atmospherical air, being able both to decompound it, and to form it a new in the most satisfactory manner. I shall, however, at present confine myself to recount such experiments as are most conclusive upon this head; and I may consider most of these as my own, having either first invented them, or having repeated those of others, with the intention of analysing atmospherical air, in perfectly new points of view.

I took a matrass (A, fig. 14. plate II.) of about 36 cubical inches capacity, having a long neck B C D E, of six or seven lines internal diameter, and having bent the neck as in Plate IV. Fig. 2. so as to allow of its being placed in[Pg 34] the furnace M M N N, in such a manner that the extremity of its neck E might be inserted under a bell-glass F G, placed in a trough of quicksilver R R S S; I introduced four ounces of pure mercury into the matrass, and, by means of a syphon, exhausted the air in the receiver F G, so as to raise the quicksilver to L L, and I carefully marked the height at which it stood by pasting on a slip of paper. Having accurately noted the height of the thermometer and barometer, I lighted a fire in the furnace M M N N, which I kept up almost continually during twelve days, so as to keep the quicksilver always almost at its boiling point. Nothing remarkable took place during the first day: The Mercury, though not boiling, was continually evaporating, and covered the interior surface of the vessels with small drops, at first very minute, which gradually augmenting to a sufficient size, fell back into the mass at the bottom of the vessel. On the second day, small red particles began to appear on the surface of the mercury, which, during the four or five following days, gradually increased in size and number; after which they ceased to increase in either respect. At the end of twelve days, seeing that the calcination of the mercury did not at all increase, I extinguished the fire, and allowed the vessels to cool. The bulk of air in the body and neck of the matrass, and in the bell-glass, reduced to[Pg 35] a medium of 28 inches of the barometer and 10° (54.5°) of the thermometer, at the commencement of the experiment was about 50 cubical inches. At the end of the experiment the remaining air, reduced to the same medium pressure and temperature, was only between 42 and 43 cubical inches; consequently it had lost about 1/6 of its bulk. Afterwards, having collected all the red particles, formed during the experiment, from the running mercury in which they floated, I found these to amount to 45 grains.

I was obliged to repeat this experiment several times, as it is difficult in one experiment both to preserve the whole air upon which we operate, and to collect the whole of the red particles, or calx of mercury, which is formed during the calcination. It will often happen in the sequel, that I shall, in this manner, give in one detail the results of two or three experiments of the same nature.

The air which remained after the calcination of the mercury in this experiment, and which was reduced to 5/6 of its former bulk, was no longer fit either for respiration or for combustion; animals being introduced into it were suffocated in a few seconds, and when a taper was plunged into it, it was extinguished as if it had been immersed into water.[Pg 36]

In the next place, I took the 45 grains of red matter formed during this experiment, which I put into a small glass retort, having a proper apparatus for receiving such liquid, or gasseous product, as might be extracted: Having applied a fire to the retort in a furnace, I observed that, in proportion as the red matter became heated, the intensity of its colour augmented. When the retort was almost red hot, the red matter began gradually to decrease in bulk, and in a few minutes after it disappeared altogether; at the same time 41-1/2 grains of running mercury were collected in the recipient, and 7 or 8 cubical inches of elastic fluid, greatly more capable of supporting both respiration and combustion than atmospherical air, were collected in the bell-glass.

A part of this air being put into a glass tube of about an inch diameter, showed the following properties: A taper burned in it with a dazzling splendour, and charcoal, instead of consuming quietly as it does in common air, burnt with a flame, attended with a decrepitating noise, like phosphorus, and threw out such a brilliant light that the eyes could hardly endure it. This species of air was discovered almost at the same time by Mr Priestley, Mr Scheele, and myself. Mr Priestley gave it the name of *dephlogisticated air*, Mr Scheele called it *empyreal air*. At first I named it *highly respirable air*, to[Pg 37] which has since been substituted the term of *vital air*. We shall presently see what we ought to think of these denominations.

In reflecting upon the circumstances of this experiment, we readily perceive, that the mercury, during its calcination, absorbs the salubrious and respirable part of the air, or, to speak more strictly, the base of this respirable part; that the remaining air is a species of mephitis, incapable of supporting combustion or respiration; and consequently that atmospheric air is composed of two elastic fluids of different and opposite qualities. As a proof of this important truth, if we recombine these two elastic fluids, which we have separately obtained in the above experiment, viz. the 42 cubical inches of mephitis, with the 8 cubical inches of respirable air, we reproduce an air precisely similar to that of the atmosphere, and possessing nearly the same power of supporting combustion and respiration, and of contributing to the calcination of metals.

Although this experiment furnishes us with a very simple means of obtaining the two principal elastic fluids which compose our atmosphere, separate from each other, yet it does not give us an exact idea of the proportion in which these two enter into its composition: For the attraction of mercury to the respirable part of the air, or rather to its base, is not sufficiently strong to overcome all the circumstances which[Pg 38] oppose this union. These obstacles are the mutual adhesion of the two constituent parts of the atmosphere for each other, and the elective attraction which unites the base of vital air with caloric; in consequence of these, when the calcination ends, or is at least carried as far as possible, in a determinate quantity of atmospheric air, there still remains a portion of respirable air united to the mephitis, which the mercury cannot separate. I shall afterwards show, that, at least in our climate, the atmospheric air is composed of respirable and mephitic airs, in the proportion of 27 and 73; and I shall then discuss the causes of the uncertainty which still exists with respect to the exactness of that proportion.

Since, during the calcination of mercury, air is decomposed, and the base of its respirable part is fixed and combined with the mercury, it follows, from the principles already established, that caloric and light must be disengaged during the process: But the two following causes prevent us from being sensible of this taking place: As the calcination lasts during several days, the disengagement of caloric and light, spread out in a considerable space of time, becomes extremely small for each particular moment of that time, so as not to be perceptible; and, in the next place, the operation being carried on by means of fire in a furnace, the heat[Pg 39] produced by the calcination itself becomes confounded with that proceeding from the furnace. I might add the respirable part of the air, or rather its base, in entering into combination with the mercury, does not part with all the caloric which it contained, but still retains a part of it after forming the new compound; but the discussion of this point, and its proofs from experiment, do not belong to this part of our subject.

It is, however, easy to render this disengagement of caloric and light evident to the senses, by causing the decomposition of air to take place in a more rapid manner. And for this purpose, iron is excellently adapted, as it possesses a much stronger affinity for the base of respirable air than mercury. The elegant experiment of Mr Ingenhouz, upon the combustion of iron, is well known. Take a piece of fine iron wire, twisted into a spiral, (BC, Plate IV. Fig. 17.) fix one of its extremities B into the cork A, adapted to the neck of the bottle DEFG, and fix to the other extremity of the wire C, a small morsel of tinder. Matters being thus prepared, fill the bottle DEFG with air deprived of its mephitic part; then light the tinder, and introduce it quickly with the wire upon which it is fixed, into the bottle which you stop up with the cork A, as is shown in the figure (17 Plate IV.) The instant the[Pg 40] tinder comes into contact with the vital air it begins to burn with great intensity; and, communicating the inflammation to the iron-wire, it too takes fire, and burns rapidly, throwing out brilliant sparks, which fall to the bottom of the vessel in rounded globules, which become black in cooling, but retain a degree of metallic splendour. The iron thus burnt is more brittle even than glass, and is easily reduced into powder, and is still attractable by the magnet, though not so powerfully as it was before combustion. As Mr Ingenhouz has neither examined the change produced on iron, nor upon the air by this operation, I have repeated the experiment under different circumstances, in an apparatus adapted to answer my particular views, as follows.

Having filled a bell-glass (A, Plate IV. Fig. 3.) of about six pints measure, with pure air, or the highly respirable part of air, I transported this jar by means of a very flat vessel, into a quicksilver bath in the bason BC, and I took care to render the surface of the mercury perfectly dry both within and without the jar with blotting paper. I then provided a small capsule of china-ware D, very flat and open, in which I placed some small pieces of iron, turned spirally, and arranged in such a way as seemed most favourable for the combustion being communicated to every part. To the end of one of these pieces of iron was[Pg 41] fixed a small morsel of tinder, to which was added about the sixteenth part of a grain of phosphorus, and, by raising the bell-glass a little, the china capsule, with its contents, were introduced into the pure air. I know that, by this means, some common air must mix with the pure air in the glass; but this, when it is done dexterously, is so very trifling, as not to injure the success of the experiment. This being done, a part of the air is sucked out from the bell-glass, by means of a syphon GHI, so as to raise the mercury within the glass to EF; and, to prevent the mercury from getting into the syphon, a small piece of paper is twisted round its extremity. In sucking out the air, if the motion of the lungs only be used, we cannot make the mercury rise above an inch or an inch and a half; but, by properly using the muscles of the mouth, we can, without difficulty, cause it to rise six or seven inches.

I next took an iron wire, (MN, Plate IV. Fig. 16.) properly bent for the purpose, and making it red hot in the fire, passed it through the mercury into the receiver, and brought it in contact with the small piece of phosphorus attached to the tinder. The phosphorus instantly takes fire, which communicates to the tinder, and from that to the iron. When the pieces have been properly arranged, the whole iron burns, even to the last particle,[Pg 42] throwing out a white brilliant light similar to that of Chinese fireworks. The great heat produced by this combustion melts the iron into round globules of different sizes, most of which fall into the China cup; but some are thrown out of it, and swim upon the surface of the mercury. At the beginning of the combustion, there is a slight augmentation in the volume of the air in the bell-glass, from the dilatation caused by the heat; but, presently afterwards, a rapid diminution of the air takes place, and the mercury rises in the glass; insomuch that, when the quantity of iron is sufficient, and the air operated upon is very pure, almost the whole air employed is absorbed.

It is proper to remark in this place, that, unless in making experiments for the purpose of discovery, it is better to be contented with burning a moderate quantity of iron; for, when this experiment is pushed too far, so as to absorb much of the air, the cup D, which floats upon the quicksilver, approaches too near the bottom of the bell-glass; and the great heat produced, which is followed by a very sudden cooling, occasioned by the contact of the cold mercury, is apt to break the glass. In which case, the sudden fall of the column of mercury, which happens the moment the least flaw is produced in the glass, causes such a wave, as throws a great part of the quicksilver from the bason. To avoid[Pg 43] this inconvenience, and to ensure success to the experiment, one gross and a half of iron is sufficient to burn in a bell-glass, which holds about eight pints of air. The glass ought likewise to be strong, that it may be able to bear the weight of the column of mercury which it has to support.

By this experiment, it is not possible to determine, at one time, both the additional weight acquired by the iron, and the changes which have taken place in the air. If it is wished to ascertain what additional weight has been gained by the iron, and the proportion between that and the air absorbed, we must carefully mark upon the bell-glass, with a diamond, the height of the mercury, both before and after the experiment[2]. After this, the syphon (GH, Pl. IV. fig. 3.) guarded, as before, with a bit of paper, to prevent its filling with mercury, is to be introduced under the bell-glass, having the thumb placed upon the extremity, G, of the syphon, to regulate the passage of the air; and by this means the air is gradually admitted, so as to let the mercury fall to its level. This being done, the bell-glass is to be carefully removed, the[Pg 44] globules of melted iron contained in the cup, and those which have been scattered about, and swim upon the mercury, are to be accurately collected, and the whole is to be weighed. The iron will be found in that state called *martial ethiops* by the old chemists, possessing a degree of metallic brilliancy, very friable, and readily reducible into powder, under the hammer, or with a pestle and mortar. If the experiment has succeeded well, from 100 grains of iron will be obtained 135 or 136 grains of ethiops, which is an augmentation of 35 per cent.

If all the attention has been paid to this experiment which it deserves, the air will be found diminished in weight exactly equal to what the iron has gained. Having therefore burnt 100 grains of iron, which has acquired an additional weight of 35 grains, the diminution of air will be found exactly 70 cubical inches; and it will be found, in the sequel, that the weight of vital air is pretty nearly half a grain for each cubical inch; so that, in effect, the augmentation of weight in the one exactly coincides with the loss of it in the other.

I shall observe here, once for all, that, in every experiment of this kind, the pressure and temperature of the air, both before and after the experiment, must be reduced, by calculation, to a common standard of 10° (54.5°) of the thermometer, and 28 inches of the barometer.[Pg 45] Towards the end of this work, the manner of performing this very necessary reduction will be found accurately detailed.

If it be required to examine the nature of the air which remains after this experiment, we must operate in a somewhat different manner. After the combustion is finished, and the vessels have cooled, we first take out the cup, and the burnt iron, by introducing the hand through the quicksilver, under the bell-glass; we next introduce some solution of potash, or caustic alkali, or of the sulphuret of potash, or such other substance as is judged proper for examining their action upon the residuum of air. I shall, in the sequel, give an account of these methods of analysing air, when I have explained the nature of these different substances, which are only here in a manner accidentally mentioned. After this examination, so much water must be let into the glass as will displace the quicksilver, and then, by means of a shallow dish placed below the bell-glass, it is to be removed into the common water pneumato-chemical apparatus, where the air remaining may be examined at large, and with great facility.

When very soft and very pure iron has been employed in this experiment, and, if the combustion has been performed in the purest respirable or vital air, free from all admixture of the noxious or mephitic part, the air which remains[Pg 46] after the combustion will be found as pure as it was before; but it is difficult to find iron entirely free from a small portion of charry matter, which is chiefly abundant in steel. It is likewise exceedingly difficult to procure the pure air perfectly free from some admixture of mephitis, with which it is almost always contaminated; but this species of noxious air does not, in the smallest degree, disturb the result of the experiment, as it is always found at the end exactly in the same proportion as at the beginning.

I mentioned before, that we have two ways of determining the constituent parts of atmospheric air, the method of analysis, and that by synthesis. The calcination of mercury has furnished us with an example of each of these methods, since, after having robbed the respirable part of its base, by means of the mercury, we have restored it, so as to recompose an air precisely similar to that of the atmosphere. But we can equally accomplish this synthetic composition of atmospheric air, by borrowing the materials of which it is composed from different kingdoms of nature. We shall see hereafter that, when animal substances are dissolved in the nitric acid, a great quantity of gas is disengaged, which extinguishes light, and is unfit for animal respiration, being exactly similar to the noxious or mephitic part of atmospheric air. And, if we take 73 parts, by weight, of this elastic[Pg 47] fluid, and mix it with 27 parts of highly respirable air, procured from calcined mercury, we will form an elastic fluid precisely similar to atmospheric air in all its properties.

There are many other methods of separating the respirable from the noxious part of the atmospheric air, which cannot be taken notice of in this part, without anticipating information, which properly belongs to the subsequent chapters. The experiments already adduced may suffice for an elementary treatise; and, in matters of this nature, the choice of our evidences is of far greater consequence than their number.

I shall close this article, by pointing out the property which atmospheric air, and all the known gasses, possess of dissolving water, which is of great consequence to be attended to in all experiments of this nature. Mr Saussure found, by experiment, that a cubical foot of atmospheric air is capable of holding 12 grains of water in solution: Other gasses, as the carbonic acid, appear capable of dissolving a greater quantity; but experiments are still wanting by which to determine their several proportions. This water, held in solution by gasses, gives rise to particular phenomena in many experiments, which require great attention, and which has frequently proved the source of great errors to chemists in determining the results of their experiments.

FOOTNOTES:

[9] It will likewise be necessary to take care that the air contained in the glass, both before and after the experiment, be reduced to a common temperature and pressure, otherwise the results of the following calculations will be fallacious.—E.

CHAP. IV.

[Pg 48]
Nomenclature of the several Constituent Parts of Atmospheric Air.

Hitherto I have been obliged to make use of circumlocution, to express the nature of the several substances which constitute our atmosphere, having provisionally used the terms of *respirable* and *noxious*, or *non-respirable parts of the air*. But the investigations I mean to undertake require a more direct mode of expression; and, having now endeavoured to give simple and distinct ideas of the different substances which enter into the composition of the atmosphere, I shall henceforth express these ideas by words equally simple.

The temperature of our earth being very near to that at which water becomes solid, and reciprocally changes from solid to fluid, and as this phenomenon takes place frequently under our observation, it has very naturally followed, that, in the languages of at least every climate subjected to any degree of winter, a term has been used for signifying water in the state of solidity, when deprived of its caloric. The same, however, has not been found necessary[Pg 49] with respect to water reduced to the state of vapour by an additional dose of caloric; since those persons who do not make a particular study of objects of this kind, are still ignorant that water, when in a temperature only a little above the boiling heat, is changed into an elastic aëriform fluid, susceptible, like all other gasses, of being received and contained in vessels, and preserving its gasseous form so long as it remains at the temperature of 80° (212°), and under a pressure not exceeding 28 inches of the mercurial barometer. As this phenomenon has not been generally observed, no language has used a particular term for expressing water in this state[10]; and the same thing occurs with all fluids, and all substances, which do not evaporate in the common temperature, and under the usual pressure of our atmosphere.

For similar reasons, names have not been given to the liquid or concrete states of most of the aëriform fluids: These were not known to arise from the combination of caloric with certain bases; and, as they had not been seen either in the liquid or solid states, their existence, under these forms, was even unknown to natural philosophers.[Pg 50]

We have not pretended to make any alteration upon such terms as are sanctified by ancient custom; and, therefore, continue to use the words *water* and *ice* in their common acceptation: We likewise retain the word *air*, to express that collection of elastic fluids which composes our atmosphere; but we have not thought it necessary to preserve the same respect for modern terms, adopted by latter philosophers, having considered ourselves as at liberty to reject such as appeared liable to occasion erroneous ideas of the substances they are meant to express, and either to substitute new terms, or to employ the old ones, after modifying them in such a manner as to convey more determinate ideas. New words have been drawn, chiefly from the Greek language, in such a manner as to make their etymology convey some idea of what was meant to be represented; and these we have always endeavoured to make short, and of such a nature as to be changeable into adjectives and verbs.

Following these principles, we have, after Mr Macquer's example, retained the term*gas*, employed by Vanhelmont, having arranged the numerous class of elastic aëriform fluids under that name, excepting only atmospheric air. *Gas*, therefore, in our nomenclature, becomes a generic term, expressing the fullest degree of saturation in any body with caloric; being, in[Pg 51] fact, a term expressive of a mode of existence. To distinguish each species of gas, we employ a second term from the name of the base, which, saturated with caloric, forms each particular gas. Thus, we name water combined to saturation with caloric, so as to form an elastic fluid, *aqueous gas*; ether, combined in the same manner, *etherial gas*; the combination of alkohol with caloric, becomes *alkoholic gas*; and, following the same principles, we have *muriatic acid gas*, *ammoniacal gas*, and so on of every substance susceptible of being combined with caloric, in such a manner as to assume the gasseous or elastic aëriform state.

We have already seen, that the atmospheric air is composed of two gasses, or aëriform fluids, one of which is capable, by respiration, of contributing to animal life, and in which metals are calcinable, and combustible bodies may burn; the other, on the contrary, is endowed with directly opposite qualities; it cannot be breathed by animals, neither will it admit of the combustion of inflammable bodies, nor of the calcination of metals. We have given to the base of the former, or respirable portion of the air, the name of *oxygen*, from οξυς *acidum*, and γεινομας, *gignor*; because, in reality, one of the most general properties of this base is to form acids, by combining with many different substances. The union of this base with caloric[Pg 52] we term *oxygen gas*, which is the same with what was formerly called *pure*, or *vital air*. The weight of this gas, at the temperature of 10° (54.50), and under a pressure equal to 28 inches of the barometer, is half a grain for each cubical inch, or one ounce and a half to each cubical foot.

The chemical properties of the noxious portion of atmospheric air being hitherto but little known, we have been satisfied to derive the name of its base from its known quality of killing such animals as are forced to breathe it, giving it the name of *azote*, from the Greek privitive particle α and ξαη, vita; hence the name of the noxious part of atmospheric air is *azotic gas*; the weight of which, in the same temperature, and under the same pressure, is

1 *oz*. 2 *gros*. and 48 *grs*. to the cubical foot, or 0.4444 of a grain to the cubical inch. We cannot deny that this name appears somewhat extraordinary; but this must be the case with all new terms, which cannot be expected to become familiar until they have been some time in use. We long endeavoured to find a more proper designation without success; it was at first proposed to call it*alkaligen gas*, as, from the experiments of Mr Berthollet, it appears to enter into the composition of ammoniac, or volatile alkali; but then, we have as yet no proof of its making one of the constituent elements of[Pg 53] the other alkalies; beside, it is proved to compose a part of the nitric acid, which gives as good reason to have called it*nitrigen*. For these reasons, finding it necessary to reject any name upon systematic principles, we have considered that we run no risk of mistake in adopting the terms of*azote*, and *azotic gas*, which only express a matter of fact, or that property which it possesses, of depriving such animals as breathe it of their lives.

I should anticipate subjects more properly reserved for the subsequent chapters, were I in this place to enter upon the nomenclature of the several species of gasses: It is sufficient, in this part of the work, to establish the principles upon which their denominations are founded. The principal merit of the nomenclature we have adopted is, that, when once the simple elementary substance is distinguished by an appropriate term, the names of all its compounds derive readily, and necessarily, from this first denomination.

FOOTNOTES:

[10]In English, the word *steam* is exclusively appropriated to water in the state of vapour. E.

[Pg 54]

CHAP. V.
Of the Decomposition of Oxygen Gas by Sulphur, Phosphorus, and Charcoal— and of the Formation of Acids in general.

In performing experiments, it is a necessary principle, which ought never to be deviated from, that they be simplified as much as possible, and that every circumstance capable of rendering their results complicated be carefully removed. Wherefore, in the experiments which form the object of this chapter, we have never employed atmospheric air, which is not a simple substance. It is true, that the azotic gas, which forms a part of its mixture, appears to be merely passive during combustion and calcination; but, besides that it retards these operations very considerably, we are not certain but it may even alter their results in some circumstances; for which reason, I have thought it necessary to remove even this possible cause of doubt, by only making use of pure oxygen gas in the following experiments, which show the effects produced by combustion in that gas; and I shall advert to such differences as take place in the results of these, when the oxygen gas, or pure[Pg 55] vital air, is mixed, in different proportions, with azotic gas.

Having filled a bell-glass (A. Pl. iv. fig. 3), of between five and six pints measure, with oxygen gas, I removed it from the water trough, where it was filled, into the quicksilver bath, by means of a shallow glass dish slipped underneath, and having dried the mercury, I introduced 61-1/4 grains of Kunkel's phosphorus in two little China cups, like that represented at D, fig. 3. under the glass A; and that I might set fire to each of the portions of phosphorus separately, and to prevent the one from catching fire from the other, one of the dishes was covered with a piece of flat glass. I next raised the quicksilver in the bell-glass up to E F, by sucking out a sufficient portion of the gas by means of the syphon G H I. After this, by means of the crooked iron wire (fig. 16.), made red hot, I set fire to the two portions of phosphorus successively, first burning that portion which was not covered with the piece of glass. The combustion was extremely rapid, attended with a very brilliant flame, and considerable disengagement of light and heat. In consequence of the great heat induced, the gas was at first much dilated, but soon after the mercury returned to its level, and a considerable absorption of gas took place; at the same time, the[Pg 56] whole inside of the glass became covered with white light flakes of concrete phosphoric acid.

At the beginning of the experiment, the quantity of oxygen gas, reduced, as above directed, to a common standard, amounted to 162 cubical inches; and, after the combustion was finished, only 23-1/4 cubical inches, likewise reduced to the standard, remained; so that the quantity of oxygen gas absorbed during the combustion was 138-3/4 cubical inches, equal to 69.375 grains.

A part of the phosphorus remained unconsumed in the bottom of the cups, which being washed on purpose to separate the acid, weighed about 16-1/4 grains; so that about 45 grains of phosphorus had been burned: But, as it is hardly possible to avoid an error of one or two grains, I leave the quantity so far qualified. Hence, as nearly 45 grains of phosphorus had, in this experiment, united with 69.375 grains of oxygen, and as no gravitating matter could have escaped through the glass, we have a right to conclude, that the weight of the substance resulting from the combustion in form of white flakes, must equal that of the phosphorus and oxygen employed, which amounts to 114.375 grains. And we shall presently find, that these flakes consisted entirely of a solid or concrete acid. When we reduce these weights to hundredth parts, it will be found, that 100 parts of[Pg 57] phosphorus require 154 parts of oxygen for saturation, and that this combination will produce 254 parts of concrete phosphoric acid, in form of white fleecy flakes.

This experiment proves, in the most convincing manner, that, at a certain degree of temperature, oxygen possesses a stronger elective attraction, or affinity, for phosphorus than for caloric; that, in consequence of this, the phosphorus attracts the base of oxygen gas from the caloric, which, being set free, spreads itself over the surrounding bodies. But, though this experiment be so far perfectly conclusive, it is not sufficiently rigorous, as, in the apparatus described, it is impossible to ascertain the weight of the flakes of concrete acid which are formed; we can therefore only determine this by calculating the weights of oxygen and phosphorus employed; but as, in physics, and in chemistry, it is not allowable to suppose what is capable of being ascertained by direct experiment, I thought it necessary to rep at this experiment, as follows, upon a larger scale, and by means of a different apparatus.

I took a large glass baloon (A. Pl. iv. fig. 4.) with an opening three inches diameter, to which was fitted a crystal stopper ground with emery, and pierced with two holes for the tubes yyy, xxx. Before shutting the baloon with its stopper, I introduced the support BC, surmounted[Pg 58] by the china cup D, containing 150 *grs.* of phosphorus; the stopper was then fitted to the opening of the baloon, luted with fat lute, and covered with slips of linen spread with quick-lime and white of eggs: When the lute was perfectly dry, the weight of the whole apparatus was determined to within a grain, or a grain and a half. I next exhausted the baloon, by means of an air pump applied to the tube XXX, and then introduced oxygen gas by means of the tube yyy, having a stop cock adapted to it. This kind of experiment is most readily and most exactly performed by means of the hydro-pneumatic machine described by Mr Meusnier and me in the Memoirs of the Academy for 1782, pag. 466. and explained in the latter part of this work, with several important additions and corrections since made to it by Mr Meusnier. With this instrument we can readily ascertain, in the most exact manner, both the quantity of oxygen gas introduced into the baloon, and the quantity consumed during the course of the experiment.

When all things were properly disposed, I set fire to the phosphorus with a burning glass. The combustion was extremely rapid, accompanied with a bright flame, and much heat; as the operation went on, large quantities of white flakes attached themselves to the inner surface of the baloon, so that at last it was rendered[Pg 59] quite opake. The quantity of these flakes at last became so abundant, that, although fresh oxygen gas was continually supplied, which ought to have supported the combustion, yet the phosphorus was soon extinguished. Having allowed the apparatus to cool completely, I first ascertained the quantity of oxygen gas employed, and weighed the baloon accurately, before it was opened. I next washed, dried, and weighed the small quantity of phosphorus remaining in the cup, on purpose to determine the whole quantity of phosphorus consumed in the experiment; this residuum of the phosphorus was of a yellow ochrey colour. It is evident, that by these several precautions, I could easily determine, 1st, the weight of the phosphorus consumed; 2d, the weight of the flakes produced by the combustion; and, 3d, the weight of the oxygen which had combined with the phosphorus. This experiment gave very nearly the same results with the former, as it proved that the phosphorus, during its combustion, had absorbed a little more than one and a half its weight of oxygen; and I learned with more certainty, that the weight of the new substance, produced in the experiment, exactly equalled the sum of the weights of the phosphorus consumed, and oxygen absorbed, which indeed was easily determinable *a priori*. If the oxygen gas employed be pure, the residuum after combustion[Pg 60] is as pure as the gas employed; this proves that nothing escapes from the phosphorus, capable of altering the purity of the oxygen gas, and that the only action of the phosphorus is to separate the oxygen from the caloric, with which it was before united.

I mentioned above, that when any combustible body is burnt in a hollow sphere of ice, or in an apparatus properly constructed upon that principle, the quantity of ice melted during the combustion is an exact measure of the quantity of caloric disengaged. Upon this head, the memoir given by M. de la Place and me, A°. 1780, p. 355, may be consulted. Having submitted the combustion of phosphorus to this trial, we found that one pound of phosphorus melted a little more than 100 pounds of ice during its combustion.

The combustion of phosphorus succeeds equally well in atmospheric air as in oxygen gas, with this difference, that the combustion is vastly slower, being retarded by the large proportion of azotic gas mixed with the oxygen gas, and that only about one-fifth part of the air employed is absorbed, because as the oxygen gas only is absorbed, the proportion of the azotic gas becomes so great toward the close of the experiment, as to put an end to the combustion.[Pg 61]

I have already shown, that phosphorus is changed by combustion into an extremely light, white, flakey matter; and its properties are entirely altered by this transformation: From being insoluble in water, it becomes not only soluble, but so greedy of moisture, as to attract the humidity of the air with astonishing rapidity; by this means it is converted into a liquid, considerably more dense, and of more specific gravity than water. In the state of phosphorus before combustion, it had scarcely any sensible taste, by its union with oxygen it acquires an extremely sharp and sour taste: in a word, from one of the class of combustible bodies, it is changed into an incombustible substance, and becomes one of those bodies called acids.

This property of a combustible substance to be converted into an acid, by the addition of oxygen, we shall presently find belongs to a great number of bodies: Wherefore, strict logic requires that we should adopt a common term for indicating all these operations which produce analogous results; this is the true way to simplify the study of science, as it would be quite impossible to bear all its specifical details in the memory, if they were not classically arranged. For this reason, we shall distinguish this conversion of phosphorus into an acid, by its union with oxygen, and in general every combination of oxygen with a combustible substance,[Pg 62] by the term of *oxygenation*: from which I shall adopt the verb to *oxygenate*, and of consequence shall say, that in *oxygenating* phosphorus we convert it into an acid.

Sulphur is likewise a combustible body, or, in other words, it is a body which possesses the power of decomposing oxygen gas, by attracting the oxygen from the caloric with which it was combined. This can very easily be proved, by means of experiments quite similar to those we have given with phosphorus; but it is necessary to premise, that in these operations with sulphur, the same accuracy of result is not to be expected as with phosphorus; because the acid which is formed by the combustion of sulphur is difficultly condensible, and because sulphur burns with more difficulty, and is soluble in the different gasses. But I can safely assert, from my own experiments, that sulphur in burning absorbs oxygen gas; that the resulting acid is considerably heavier than the sulphur burnt; that its weight is equal to the sum of the weights of the sulphur which has been burnt, and of the oxygen absorbed; and, lastly that this acid is weighty, incombustible, and miscible with water in all proportions: The only uncertainty remaining upon this head, is with regard to the proportions of sulphur and of oxygen which enter into the composition of the acid.[Pg 63]

Charcoal, which, from all our present knowledge regarding it, must be considered as a simple combustible body, has likewise the property of decomposing oxygen gas, by absorbing its base from the caloric: But the acid resulting from this combustion does not condense in the common temperature; under the pressure of our atmosphere, it remains in the state of gas, and requires a large proportion of water to combine with or be dissolved in. This acid has, however, all the known properties of other acids, though in a weaker degree, and combines, like them, with all the bases which are susceptible of forming neutral salts.

The combustion of charcoal in oxygen gas, may be effected like that of phosphorus in the bell-glass, (A. Pl. IV. fig. 3.) placed over mercury: but, as the heat of red hot iron is not sufficient to set fire to the charcoal, we must add a small morsel of tinder, with a minute particle of phosphorus, in the same manner as directed in the experiment for the combustion of iron. A detailed account of this experiment will be found in the memoirs of the academy for 1781, p. 448. By that experiment it appears, that 28 parts by weight of charcoal require 72 parts of oxygen for saturation, and that the aëriform acid produced is precisely equal in weight to the sum of the weights of the charcoal and oxygen gas employed.[Pg 64] This aëriform acid was called fixed or fixable air by the chemists who first discovered it; they did not then know whether it was air resembling that of the atmosphere, or some other elastic fluid, vitiated and corrupted by combustion; but since it is now ascertained to be an acid, formed like all others by the oxygenation of its peculiar base, it is obvious that the name of fixed air is quite ineligible[11].

By burning charcoal in the apparatus mentioned p. 60, Mr de la Place and I found that one *lib.* of charcoal melted 96 *libs.* 6 *oz.* of ice; that, during the combustion, 2 *libs.* 9*oz.* 1 *gros.* 10 *grs.* of oxygen were absorbed, and that 3 *libs.* 9 *oz.* 1 *gros.* 10 *grs.* of acid gas were formed. This gas weighs 0.695 parts of a grain for each cubical inch, in the common standard temperature and pressure mentioned above, so that 34,242 cubical inches of acid gas are produced by the combustion of one pound of charcoal.

I might multiply these experiments, and show by a numerous succession of facts, that all acids are formed by the combustion of certain substances; but I am prevented from doing so in[Pg 65] place, by the plan which I have laid down, of proceeding only from facts already ascertained, to such as are unknown, and of drawing my examples only from circumstances already explained. In the mean time, however, the three examples above cited may suffice for giving a clear and accurate conception of the manner in which acids are formed. By these it may be clearly seen, that oxygen is an element common to them all, which constitutes their acidity; and that they differ from each other, according to the nature of the oxygenated or acidified substance. We must therefore, in every acid, carefully distinguish between the acidifiable, base, which Mr de Morveau calls the radical, and the acidifiing principle or oxygen.

FOOTNOTES:

[11]It may be proper to remark, though here omitted by the author, that, in conformity with the general principles of the new nomenclature, this acid is by Mr Lavoisier and his coleagues called the carbonic acid, and when in the aëriform state carbonic acid gas. E.

[Pg 66]

CHAP. VI.
Of the Nomenclature of Acids in general, and particularly of those drawn from Nitre and Sea-Salt.

It becomes extremely easy, from the principles laid down in the preceding chapter, to establish a systematic nomenclature for the acids: The word *acid*, being used as a generic term, each acid falls to be distinguished in language, as in nature, by the name of its base or radical. Thus, we give the generic name of acids to the products of the combustion or oxygenation of phosphorus, of sulphur, and of charcoal; and these products are respectively named, the *phosphoric acid*, the *sulphuric acid*, and the *carbonic acid*.

There is however, a remarkable circumstance in the oxygenation of combustible bodies, and of a part of such bodies as are convertible into acids, that they are susceptible of different degrees of saturation with oxygen, and that the resulting acids, though formed by the union of the same elements, are possessed of different properties, depending upon that difference of proportion. Of this, the phosphoric acid, and more especially the sulphuric, furnishes us with examples.[Pg 67] When sulphur is combined with a small proportion of oxygen, it forms, in this first or lower degree of oxygenation, a volatile acid, having a penetrating odour, and possessed of very particular qualities. By a larger proportion of oxygen, it is changed into a fixed, heavy acid, without any odour, and which, by combination with other bodies, gives products quite different from those furnished by the former. In this instance, the principles of our nomenclature seem to fail; and it seems difficult to derive such terms from the name of the acidifiable base, as shall distinctly express these two degrees of saturation, or oxygenation, without circumlocution. By reflection, however, upon the subject, or perhaps rather from the necessity of the case, we have thought it allowable to express these varieties in the oxygenation of the acids, by simply varying the termination of their specific names. The volatile acid produced from sulphur was anciently known to Stahl under the name of *sulphurous* acid[12]. We have[Pg 68] preserved that term for this acid from sulphur under-saturated with oxygen; and distinguish the other, or completely saturated or oxygenated acid, by the name of *sulphuric* acid. We shall therefore say, in this new chemical language, that sulphur, in combining with oxygen, is susceptible of two degrees of saturation; that the first, or lesser degree, constitutes sulphurous acid, which is volatile and penetrating; whilst the second, or higher degree of saturation, produces sulphuric acid, which is fixed and inodorous. We shall adopt this difference of termination for all the acids which assume several degrees of saturation. Hence we have a phosphorous and a phosphoric acid, an acetous and an acetic acid; and so on, for others in similar circumstances.

This part of chemical science would have been extremely simple, and the nomenclature of the acids would not have been at all perplexed, as it is now in the old nomenclature, if the base or radical of each acid had been known when the acid itself was discovered. Thus, for instance, phosphorus being a known substance before the discovery of its acid, this latter was rightly distinguished by a term drawn from the name of its acidifiable base. But when, on the contrary, an acid happened to be discovered before its base, or rather, when the acidifiable base from which it was formed remained unknown,[Pg 69] names were adopted for the two, which have not the smallest connection; and thus, not only the memory became burthened with useless appellations, but even the minds of students, nay even of experienced chemists, became filled with false ideas, which time and reflection alone is capable of eradicating. We may give an instance of this confusion with respect to the acid sulphur: The former chemists having procured this acid from the vitriol of iron, gave it the name of the vitriolic acid from the name of the substance which produced it; and they were then ignorant that the acid procured from sulphur by combustion was exactly the same.

The same thing happened with the aëriform acid formerly called *fixed air*; it not being known that this acid was the result of combining charcoal with oxygen, a variety of denominations have been given to it, not one of which conveys just ideas of its nature or origin. We have found it extremely easy to correct and modify the ancient language with respect to these acids proceeding from known bases, having converted the name of *vitriolic acid* into that of *sulphuric*, and the name of *fixed air* into that of *carbonic acid*; but it is impossible to follow this plan with the acids whose bases are still unknown; with these we have been obliged to use a contrary plan, and, instead of forming the name of the acid from that of its[Pg 70] base, have been forced to denominate the unknown base from the name of the known acid, as happens in the case of the acid which is procured from sea salt.

To disengage this acid from the alkaline base with which it is combined, we have only to pour sulphuric acid upon sea-salt, immediately a brisk effervescence takes place, white vapours arise, of a very penetrating odour, and, by only gently heating the mixture, all the acid is driven off. As, in the common temperature and pressure of our atmosphere, this acid is naturally in the state of gas, we must use particular precautions for retaining it in proper vessels. For small experiments, the most simple and most commodious apparatus consists of a small retort G, (Pl. V. Fig. 5.), into which the sea-salt is introduced, well dried[13], we then pour on some concentrated sulphuric acid, and immediately introduce the beak of the retort under little jars or bell-glasses A, (same Plate and Fig.), previously filled with quicksilver. In proportion as the acid gas is disengaged, it passes into the jar, and gets to the top of the quicksilver, which it displaces. When the disengagement[Pg 71] of the gas slackens, a gentle heat is applied to the retort, and gradually increased till nothing more passes over. This acid gas has a very strong affinity with water, which absorbs an enormous quantity of it, as is proved by introducing a very thin layer of water into the glass which contains the gas; for, in an instant, the whole acid gas disappears, and combines with the water.

This latter circumstance is taken advantage of in laboratories and manufactures, on purpose to obtain the acid of sea-salt in a liquid form; and for this purpose the apparatus (Pl. IV. Fig. 1.) is employed. It consists, 1st, of a tubulated retort A, into which the sea-salt, and after it the sulphuric acid, are introduced through the opening H; 2d, of the baloon or recipient *c, b*, intended for containing the small quantity of liquid which passes over during the process; and, 3d, of a set of bottles, with two mouths, L, L, L, L, half filled with water, intended for absorbing the gas disengaged by the distillation. This apparatus will be more amply described in the latter part of this work.

Although we have not yet been able, either to compose or to decompound this acid of sea-salt, we cannot have the smallest doubt that it, like all other acids, is composed by the union of oxygen with an acidifiable base. We have therefore called this unknown substance the[Pg 72] *muriatic base*, or *muriatic radical*, deriving this name, after the example of Mr Bergman and Mr de Morveau, from the Latin word *muria*, which was anciently used to signify sea-salt. Thus, without being able exactly to determine the component parts of *muriatic acid*, we design, by that term, a volatile acid, which retains the form of gas in the common temperature and pressure of our atmosphere, which combines with great facility, and in great quantity, with water, and whose acidifiable base adheres so very intimately with oxygen, that no method has hitherto been devised for separating them. If ever this acidifiable base of the muriatic acid is discovered to be a known substance, though now unknown in that capacity, it will be requisite to change its present denomination for one analogous with that of its base.

In common with sulphuric acid, and several other acids, the muriatic is capable of different degrees of oxygenation; but the excess of oxygen produces quite contrary effects upon it from what the same circumstance produces upon the acid of sulphur. The lower degree of oxygenation converts sulphur into a volatile gasseous acid, which only mixes in small proportions with water, whilst a higher oxygenation forms an acid possessing much stronger acid properties, which is very fixed and cannot remain in the state of gas but in a very high temperature, which has[Pg 73] no smell, and which mixes in large proportion with water. With muriatic acid, the direct reverse takes place; an additional saturation with oxygen renders it more volatile, of a more penetrating odour, less miscible with water, and diminishes its acid properties. We were at first inclined to have denominated these two degrees of saturation in the same manner as we had done with the acid of sulphur, calling the less oxygenated *muriatous acid*, and that which is more saturated with oxygen *muriatic acid*: But, as this latter gives very particular results in its combinations, and as nothing analogous to it is yet known in chemistry, we have left the name of muriatic acid to the less saturated, and give the latter the more compounded appellation of *oxygenated muriatic acid*.

Although the base or radical of the acid which is extracted from nitre or saltpetre be better known, we have judged proper only to modify its name in the same manner with that of the muriatic acid. It is drawn from nitre, by the intervention of sulphuric acid, by a process similar to that described for extracting the muriatic acid, and by means of the same apparatus (Pl. IV. Fig. 1.). In proportion as the acid passes over, it is in part condensed in the baloon or recipient, and the rest is absorbed by the water contained in the bottles L,L,L,L; the water becomes first green,[Pg 74] then blue, and at last yellow, in proportion to the concentration of the acid. During this operation, a large quantity of oxygen gas, mixed with a small proportion of azotic gas, is disengaged.

This acid, like all others, is composed of oxygen, united to an acidifiable base, and is even the first acid in which the existence of oxygen was well ascertained. Its two constituent elements are but weakly united, and are easily separated, by presenting any substance with which oxygen has a stronger affinity than with the acidifiable base peculiar to this acid. By some experiments of this kind, it was first discovered that azote, or the base of mephitis or azotic gas, constituted its acidifiable base or radical; and consequently that the acid of nitre was really an azotic acid, having azote for its base, combined with oxygen. For these reasons, that we might be consistent with our principles, it appeared necessary, either to call the acid by the name of *azotic*, or to name the base *nitric radical*; but from either of these we were dissuaded, by the following considerations. In the *first* place, it seemed difficult to change the name of nitre or saltpetre, which has been universally adopted in society, in manufactures, and in chemistry; and, on the other hand, azote having been discovered by Mr Berthollet to be the base of volatile alkali, or ammoniac, as well as of this acid,[Pg 75] we thought it improper to call it nitric radical. We have therefore continued the term of azote to the base of that part of atmospheric air which is likewise the nitric and ammoniacal radical; and we have named the acid of nitre, in its lower and higher degrees of oxygenation, *nitrous acid* in the former, and *nitric acid* in the latter state; thus preserving its former appellation properly modified.

Several very respectable chemists have disapproved of this deference for the old terms, and wished us to have persevered in perfecting a new chemical language, without paying any respect for ancient usage; so that, by thus steering a kind of middle course, we have exposed ourselves to the censures of one sect of chemists, and to the expostulations of the opposite party.

The acid of nitre is susceptible of assuming a great number of separate states, depending upon its degree of oxygenation, or upon the proportions in which azote and oxygen enter into its composition. By a first or lowest degree of oxygenation, it forms a particular species of gas, which we shall continue to name *nitrous gas*; this is composed nearly of two parts, by weight, of oxygen combined with one part of azote; and in this state it is not miscible with water. In this gas, the azote is by no means saturated with oxygen, but, on the contrary, has[Pg 76] still a very great affinity for that element, and even attracts it from atmospheric air, immediately upon getting into contact with it. This combination of nitrous gas with atmospheric air has even become one of the methods for determining the quantity of oxygen contained in air, and consequently for ascertaining its degree of salubrity.

This addition of oxygen converts the nitrous gas into a powerful acid, which has a strong affinity with water, and which is itself susceptible of various additional degrees of oxygenation. When the proportions of oxygen and azote is below three parts, by weight, of the former, to one of the latter, the acid is red coloured, and emits copious fumes. In this state, by the application of a gentle heat, it gives out nitrous gas; and we term it, in this degree of oxygenation, *nitrous acid*. When four parts, by weight, of oxygen, are combined with one part of azote, the acid is clear and colourless, more fixed in the fire than the nitrous acid, has less odour, and its constituent elements are more firmly united. This species of acid, in conformity with our principles of nomenclature, is called *nitric acid*.

Thus, nitric acid is the acid of nitre, surcharged with oxygen; nitrous acid is the acid of nitre surcharged with azote; or, what is the same thing, with nitrous gas; and this latter is[Pg 77] azote not sufficiently saturated with oxygen to possess the properties of an acid. To this degree of oxygenation, we have afterwards, in the course of this work, given the generical name of *oxyd*[14].

FOOTNOTES:

[12]The term formerly used by the English chemists for this acid was written *sulphureous*; but we have thought proper to spell it as above, that it may better conform with the similar terminations of nitrous, carbonous, &c. to be used hereafter. In general, we have used the English terminations *ic* and *ous* to translate the terms of the Author which end with *ique* and *eux*, with hardly any other alterations.—E.

[13]For this purpose, the operation called *decrepitation* is used, which consists in subjecting it to nearly a red heat, in a proper vessel, so as to evaporate all its water of crystallization.—E.

[14]In strict conformity with the principles of the new nomenclature, but which the Author has given his reasons for deviating from in this instance, the following ought to have been the terms for azote, in its several degrees of oxygenation: Azote, azotic gas, (azote combined with caloric), azotic oxyd gas, nitrous acid, and nitric acid.—E.

[Pg 78]

CHAP. VII.
Of the Decomposition of Oxygen Gas by means of Metals, and the Formation of Metallic Oxyds.

Oxygen has a stronger affinity with metals heated to a certain degree than with caloric; in consequence of which, all metallic bodies, excepting gold, silver, and platina, have the property of decomposing oxygen gas, by attracting its base from the caloric with which it was combined. We have already shown in what manner this decomposition takes place, by means of mercury and iron; having observed, that, in the case of the first, it must be considered as a kind of gradual combustion, whilst, in the latter, the combustion is extremely rapid, and attended with a brilliant flame. The use of the heat employed in these operations is to separate the particles of the metal from each other, and to diminish their attraction of cohesion or aggregation, or, what is the same thing, their mutual attraction for each other.

The absolute weight of metallic substances is augmented in proportion to the quantity of oxygen they absorb; they, at the same time, lose their metallic splendour, and are reduced into[Pg 79] an earthy pulverulent matter. In this state metals must not be considered as entirely saturated with oxygen, because their action upon this element is counterbalanced by the power of affinity between it and caloric. During the calcination of metals, the oxygen is therefore acted upon by two separate and opposite powers, that of its attraction for caloric, and that exerted by the metal, and only tends to unite with the latter in consequence of the excess of the latter over the former, which is, in general, very inconsiderable. Wherefore, when metallic substances are oxygenated in atmospheric air, or in oxygen gas, they are not converted into acids like sulphur, phosphorus, and charcoal, but are only changed into intermediate substances, which, though approaching to the nature of salts, have not acquired all the saline properties. The old chemists have affixed the name of *calx* not only to metals in this state, but to every body which has been long exposed to the action of fire without being melted. They have converted this word *calx* into a generical term, under which they confound calcareous earth, which, from a neutral salt, which it really was before calcination, has been changed by fire into an earthy alkali, by *losing* half of its weight, with metals which, by the same means, have joined themselves to a new substance, whose quantity often *exceeds* half their weight, and by which they[Pg 80] have been changed almost into the nature of acids. This mode of classifying substances of so very opposite natures, under the same generic name, would have been quite contrary to our principles of nomenclature, especially as, by retaining the above term for this state of metallic substances, we must have conveyed very false ideas of its nature. We have, therefore, laid aside the expression *metallic calx* altogether, and have substituted in its place the term *oxyd*, from the Greek word οξυς.

By this may be seen, that the language we have adopted is both copious and expressive. The first or lowest degree of oxygenation in bodies, converts them into *oxyds*; a second degree of additional oxygenation constitutes the class of acids, of which the specific names, drawn from their particular bases, terminate in *ous*, as the *nitrous* and *sulphurous* acids; the third degree of oxygenation changes these into the species of acids distinguished by the termination in ic, as the *nitric* and *sulphuric* acids; and, lastly, we can express a fourth, or highest degree of oxygenation, by adding the word *oxygenated* to the name of the acid, as has been already done with the *oxygenated muriatic* acid.

We have not confined the term *oxyd* to expressing the combinations of metals with oxygen, but have extended it to signify that first degree of oxygenation in all bodies, which,[Pg 81] without converting them into acids, causes them to approach to the nature of salts. Thus, we give the name of *oxyd of sulphur* to that soft substance into which sulphur is converted by incipient combustion; and we call the yellow matter left by phosphorus, after combustion, by the name of *oxyd of phosphorus*. In the same manner, nitrous gas, which is azote in its first degree of oxygenation, is the *oxyd of azote*. We have likewise oxyds in great numbers from the vegetable and animal kingdoms; and I shall show, in the sequel, that this new language throws great light upon all the operations of art and nature.

We have already observed, that almost all the metallic oxyds have peculiar and permanent colours. These vary not only in the different species of metals, but even according to the various degrees of oxygenation in the same metal. Hence we are under the necessity of adding two epithets to each oxyd, one of which indicates the metal *oxydated*[15], while the other indicates[Pg 82] the peculiar colour of the oxyd. Thus, we have the black oxyd of iron, the red oxyd of iron, and the yellow oxyd of iron; which expressions respectively answer to the old unmeaning terms of martial ethiops, colcothar, and rust of iron, or ochre. We have likewise the gray, yellow, and red oxyds of lead, which answer to the equally false or insignificant terms, ashes of lead, massicot, and minium.

These denominations sometimes become rather long, especially when we mean to indicate whether the metal has been oxydated in the air, by detonation with nitre, or by means of acids; but then they always convey just and accurate ideas of the corresponding object which we wish to express by their use. All this will be rendered perfectly clear and distinct by means of the tables which are added to this work.

FOOTNOTES:

[15]Here we see the word oxyd converted into the verb *to oxydate, oxydated, oxydating*, after the same manner with the derivation of the verb *to oxygenate, oxygenated, oxygenating*, from the word *oxygen*. I am not clear of the absolute necessity of this second verb here first introduced, but think, in a work of this nature, that it is the duty of the translator to neglect every other consideration for the sake of strict fidelity to the ideas of his author.—E.

[Pg 83]

CHAP. VIII.
Of the Radical Principle of Water, and of its Decomposition by Charcoal and Iron.

Until very lately, water has always been thought a simple substance, insomuch that the older chemists considered it as an element. Such it undoubtedly was to them, as they were unable to decompose it; or, at least, since the decomposition which took place daily before their eyes was entirely unnoticed. But we mean to prove, that water is by no means a simple or elementary substance. I shall not here pretend to give the history of this recent, and hitherto contested discovery, which is detailed in the Memoirs of the Academy for 1781, but shall only bring forwards the principal proofs of the decomposition and composition of water; and, I may venture to say, that these will be convincing to such as consider them impartially.

Experiment First.

Having fixed the glass tube EF, (Pl. vii. fig. 11.) of from 8 to 12 lines diameter, across a furnace, with a small inclination from E to F,[Pg 84] lute the superior extremity E to the glass retort A, containing a determinate quantity of distilled water, and to the inferior extremity F, the worm SS fixed into the neck of the doubly tubulated bottle H, which has the bent tube KK adapted to one of its openings, in such a manner as to convey such aëriform fluids or gasses as may be disengaged, during the experiment, into a proper apparatus for determining their quantity and nature.

To render the success of this experiment certain, it is necessary that the tube EF be made of well annealed and difficultly fusible glass, and that it be coated with a lute composed of clay mixed with powdered stone-ware; besides which, it must be supported about its middle by means of an iron bar passed through the furnace, lest it should soften and bend during the experiment. A tube of China-ware, or porcellain, would answer better than one of glass for this experiment, were it not difficult to procure one so entirely free from pores as to prevent the passage of air or of vapours.

When things are thus arranged, a fire is lighted in the furnace EFCD, which is supported of such a strength as to keep the tube EF red hot, but not to make it melt; and, at the same time, such a fire is kept up in the furnace VVXX, as to keep the water in the retort A continually boiling.[Pg 85]

In proportion as the water in the retort A is evaporated, it fills the tube EF, and drives out the air it contained by the tube KK; the aqueous gas formed by evaporation is condensed by cooling in the worm SS, and falls, drop by drop, into the tubulated bottle H. Having continued this operation until all the water be evaporated from the retort, and having carefully emptied all the vessels employed, we find that a quantity of water has passed over into the bottle H, exactly equal to what was before contained in the retort A, without any disengagement of gas whatsoever: So that this experiment turns out to be a simple distillation; and the result would have been exactly the same, if the water had been run from one vessel into the other, through the tube EF, without having undergone the intermediate incandescence.

Experiment Second.

The apparatus being disposed, as in the former experiment, 28 *grs.* of charcoal, broken into moderately small parts, and which has previously been exposed for a long time to a red heat in close vessels, are introduced into the tube EF. Every thing else is managed as in the preceding experiment.

The water contained in the retort A is distilled, as in the former experiment, and, being[Pg 86] condensed in the worm, falls into the bottle H; but, at the same time, a considerable quantity of gas is disengaged, which, escaping by the tube KK, is received in a convenient apparatus for that purpose. After the operation is finished, we find nothing but a few atoms of ashes remaining in the tube EF; the 28 *grs.* of charcoal having entirely disappeared.

When the disengaged gasses are carefully examined, they are sound to weigh 113.7*grs.*[16]; these are of two kinds, viz. 144 cubical inches of carbonic acid gas, weighing 100 *grs.* and 380 cubical inches of a very light gas, weighing only 13.7 *grs.* which takes fire when in contact with air, by the approach of a lighted body; and, when the water which has passed over into the bottle H is carefully examined, it is found to have lost 85.7 *grs.* of its weight. Thus, in this experiment, 85.7 *grs.* of water, joined to 28 *grs.* of charcoal, have combined in such a way as to form 100 *grs.* of carbonic acid, and 13.7 *grs.* of a particular gas capable of being burnt.

I have already shown, that 100 *grs.* of carbonic acid gas consists of 72 *grs.* of oxygen, combined with 28 *grs.* of charcoal; hence the 28[Pg 87] *grs.* of charcoal placed in the glass tube have acquired 72 *grs.* of oxygen from the water; and it follows, that 85.7 *grs.* of water are composed of 72 *grs.* of oxygen, combined with 13.7 *grs.* of a gas susceptible of combustion. We shall see presently that this gas cannot possibly have been disengaged from the charcoal, and must, consequently, have been produced from the water.

I have suppressed some circumstances in the above account of this experiment, which would only have complicated and obscured its results in the minds of the reader. For instance, the inflammable gas dissolves a very small part of the charcoal, by which means its weight is somewhat augmented, and that of the carbonic gas proportionally diminished. Altho' the alteration produced by this circumstance is very inconsiderable; yet I have thought it necessary to determine its effects by rigid calculation, and to report, as above, the results of the experiment in its simplified state, as if this circumstance had not happened. At any rate, should any doubts remain respecting the consequences I have drawn from this experiment, they will be fully dissipated by the following experiments, which I am going to adduce in support of my opinion.[Pg 88]

Experiment Third.

The apparatus being disposed exactly as in the former experiment, with this difference, that instead of the 28 *grs.* of charcoal, the tube EF is filled with 274 *grs.* of soft iron in thin plates, rolled up spirally. The tube is made red hot by means of its furnace, and the water in the retort A is kept constantly boiling till it be all evaporated, and has passed through the tube EF, so as to be condensed in the bottle H.

No carbonic acid gas is disengaged in this experiment, instead of which we obtain 416 cubical inches, or 15 *grs.* of inflammable gas, thirteen times lighter than atmospheric air. By examining the water which has been distilled, it is found to have lost 100 *grs.*and the 274 *grs.* of iron confined in the tube are found to have acquired 85 *grs.*additional weight, and its magnitude is considerably augmented. The iron is now hardly at all attractable by the magnet; it dissolves in acids without effervescence; and, in short, it is converted into a black oxyd, precisely similar to that which has been burnt in oxygen gas.

In this experiment we have a true *oxydation* of iron, by means of water, exactly similar to that produced in air by the assistance of heat. One hundred grains of water having been decomposed,[Pg 89] 85 *grs.* of oxygen have combined with the iron, so as to convert it into the state of black oxyd, and 15 *grs.* of a peculiar inflammable gas are disengaged: From all this it clearly follows, that water is composed of oxygen combined with the base of an inflammable gas, in the respective proportions of 85 parts, by weight of the former, to 15 parts of the latter.

Thus water, besides the oxygen, which is one of its elements in common with many other substances, contains another element as its constituent base or radical, and for which we must find an appropriate term. None that we could think of seemed better adapted than the word *hydrogen*, which signifies the *generative principle of water*, from ὑδορ *aqua*, and γεινομας *gignor*[17]. We call the combination of this element with caloric *hydrogen gas*; and the term hydrogen expresses the base of that gas, or the radical of water.

[Pg 90]
This experiment furnishes us with a new combustible body, or, in other words, a body which has so much affinity with oxygen as to draw it from its connection with caloric, and to decompose air or oxygen gas. This combustible body has itself so great affinity with caloric, that, unless when engaged in a combination with some other body, it always subsists in the aëriform or gasseous state, in the usual temperature and pressure of our atmosphere. In this state of gas it is about 1/13 of the weight of an equal bulk of atmospheric air; it is not absorbed by water, though it is capable of holding a small quantity of that fluid in solution, and it is incapable of being used for respiration.

As the property this gas possesses, in common with all other combustible bodies, is nothing more than the power of decomposing air, and carrying off its oxygen from the caloric with which it was combined, it is easily understood that it cannot burn, unless in contact with air or oxygen gas. Hence, when we set fire to a bottle full of this gas, it burns gently, first at the neck of the bottle, and then in the inside of it, in proportion as the external air gets in: This combustion is slow and successive, and only takes place at the surface of contact between the two gasses. It is quite different when the two gasses are mixed before they are set on fire: If, for instance, after having introduced one part of[Pg 91] oxygen gas into a narrow mouthed bottle, we fill it up with two parts of hydrogen gas, and bring a lighted taper, or other burning body, to the mouth of the bottle, the combustion of the two gasses takes place instantaneously with a violent explosion. This experiment ought only to be made in a bottle of very strong green glass, holding not more than a pint, and wrapped round with twine, otherwise the operator will be exposed to great danger from the rupture of the bottle, of which the fragments will be thrown about with great force.

If all that has been related above, concerning the decomposition of water, be exactly conformable to truth;—if, as I have endeavoured to prove, that substance be really composed of hydrogen, as its proper constituent element, combined with oxygen, it ought to follow, that, by reuniting these two elements together, we should recompose water; and that this actually happens may be judged of by the following experiment.

Experiment Fourth.

I took a large cristal baloon, A, Pl. iv. fig. 5. holding about 30 pints, having a large opening, to which was cemented the plate of copper BC, pierced with four holes, in which four tubes terminate. The first tube, H h, is intended to[Pg 92] be adapted to an air pump, by which the baloon is to be exhausted of its air. The second tube gg, communicates, by its extremity MM, with a reservoir of oxygen gas, with which the baloon is to be filled. The third tube d D d', communicates, by its extremity d NN, with a reservoir of hydrogen gas. The extremity d' of this tube terminates in a capillary opening, through which the hydrogen gas contained in the reservoir is forced, with a moderate degree of quickness, by the pressure of one or two inches of water. The fourth tube contains a metallic wire GL, having a knob at its extremity L, intended for giving an electrical spark from L to d', on purpose to set fire to the hydrogen gas: This wire is moveable in the tube, that we may be able to separate the knob L from the extremity d' of the tube D d'. The three tubes d D d', gg, and H h, are all provided with stop-cocks.

That the hydrogen gas and oxygen gas may be as much as possible deprived of water, they are made to pass, in their way to the baloon A, through the tubes MM, NN, of about an inch diameter, and filled with salts, which, from their deliquescent nature, greedily attract the moisture of the air: Such are the acetite of potash, and the muriat or nitrat of lime[18]. These salts[Pg 93] must only be reduced to a coarse powder, lest they run into lumps, and prevent the gasses from geting through their interstices.

We must be provided before hand with a sufficient quantity of oxygen gas, carefully purified from all admixture of carbonic acid, by long contact with a solution of potash[19].

We must likewise have a double quantity of hydrogen gas, carefully purified in the same manner by long contact with a solution of potash in water. The best way of obtaining this gas free from mixture is, by decomposing water with very pure soft iron, as directed in Exp. 3. of this chapter.

Having adjusted every thing properly, as above directed, the tube H h is adapted to an air-pump, and the baloon A is exhausted of its air. We next admit the oxygen gas so as to fill the baloon, and then, by means of pressure, as is before mentioned, force a small stream of hydrogen gas through its tube D d', which we immediately set on fire by an electric spark. By means of the above described apparatus, we can[Pg 94] continue the mutual combustion of these two gasses for a long time, as we have the power of supplying them to the baloon from their reservoirs, in proportion as they are consumed. I have in another place[20] given a description of the apparatus used in this experiment, and have explained the manner of ascertaining the quantities of the gasses consumed with the most scrupulous exactitude.

In proportion to the advancement of the combustion, there is a deposition of water upon the inner surface of the baloon or matrass A: The water gradually increases in quantity, and, gathering into large drops, runs down to the bottom of the vessel. It is easy to ascertain the quantity of water collected, by weighing the baloon both before and after the experiment. Thus we have a twofold verification of our experiment, by ascertaining both the quantities of the gasses employed, and of the water formed by their combustion: These two quantities must be equal to each other. By an operation of this kind, Mr Meusnier and I ascertained that it required 85 parts, by weight, of oxygen, united to 15 parts of hydrogen, to compose 100 parts of water. This experiment, which has not hitherto been published, was made in presence of a numerous committee from the Royal Academy.[Pg 95] We exerted the most scrupulous attention to its accuracy; and have reason to believe that the above propositions cannot vary a two hundredth part from absolute truth.

From these experiments, both analytical and synthetic, we may now affirm that we have ascertained, with as much certainty as is possible in physical or chemical subjects, that water is not a simple elementary substance, but is composed of two elements, oxygen and hydrogen; which elements, when existing separately, have so strong affinity for caloric, as only to subsist under the form of gas in the common temperature and pressure of our atmosphere.

This decomposition and recomposition of water is perpetually operating before our eyes, in the temperature of the atmosphere, by means of compound elective attraction. We shall presently see that the phenomena attendant upon vinous fermentation, putrefaction, and even vegetation, are produced, at least in a certain degree, by decomposition of water. It is very extraordinary that this fact should have hitherto been overlooked by natural philosophers and chemists: Indeed, it strongly proves, that, in chemistry, as in moral philosophy, it is extremely difficult to overcome prejudices imbibed in early education, and to search for truth in any other road than the one we have been accustomed to follow.[Pg 96]

I shall finish this chapter by an experiment much less demonstrative than those already related, but which has appeared to make more impression than any other upon the minds of many people. When 16 ounces of alkohol are burnt in an apparatus[21] properly adapted for collecting all the water disengaged during the combustion, we obtain from 17 to 18 ounces of water. As no substance can furnish a product larger than its original bulk, it follows, that something else has united with the alkohol during its combustion; and I have already shown that this must be oxygen, or the base of air. Thus alkohol contains hydrogen, which is one of the elements of water; and the atmospheric air contains oxygen, which is the other element necessary to the composition of water. This experiment is a new proof that water is a compound substance.

FOOTNOTES:

[16] In the latter part of this work will be found a particular account of the processes necessary for separating the different kinds of gasses, and for determining their quantities.—A.

[17]This expression Hydrogen has been very severely criticised by some, who pretend that it signifies engendered by water, and not that which engenders water. The experiments related in this chapter prove, that, when water is decomposed, hydrogen is produced, and that, when hydrogen is combined with oxygen, water is produced: So that we may say, with equal truth, that water is produced from hydrogen, or hydrogen is produced from water.—A.

[18]See the nature of these salts in the second part of this book.—A.

[19]By potash is here meant, pure or caustic alkali, deprived of carbonic acid by means of quick-lime: In general, we may observe here, that all the alkalies and earths must invariably be considered as in their pure or caustic state, unless otherwise expressed.—E. The method of obtaining this pure alkali of potash will be given in the sequel.—A.

[20]See the third part of this work.—A.

[21]See an account of this apparatus in the third part of this work.—A.

[Pg 97]

CHAP. IX.
Of the quantities of Caloric disengaged from different species of Combustion.

We have already mentioned, that, when any body is burnt in the center of a hollow sphere of ice and supplied with air at the temperature of zero (32°), the quantity of ice melted from the inside of the sphere becomes a measure of the relative quantities of caloric disengaged. Mr de la Place and I gave a description of the apparatus employed for this kind of experiment in the Memoirs of the Academy for 1780, p. 355; and a description and plate of the same apparatus will be found in the third part of this work. With this apparatus, phosphorus, charcoal, and hydrogen gas, gave the following results:

One pound of phosphorus melted 100 *libs.* of ice.

One pound of charcoal melted 96 *libs.* 8 *oz.*

One pound of hydrogen gas melted 295 *libs.* 9 *oz.* 3-1/2 *gros.*

As a concrete acid is formed by the combustion of phosphorus, it is probable that very little caloric remains in the acid, and, consequently,[Pg 98] that the above experiment gives us very nearly the whole quantity of caloric contained in the oxygen gas. Even if we suppose the phosphoric acid to contain a good deal of caloric, yet, as the phosphorus must have contained nearly an equal quantity before combustion, the error must be very small, as it will only consist of the difference between what was contained in the phosphorus before, and in the phosphoric acid after combustion.

I have already shown in Chap. V. that one pound of phosphorus absorbs one pound eight ounces of oxygen during combustion; and since, by the same operation, 100 *lib.*of ice are melted, it follows, that the quantity of caloric contained in one pound of oxygen gas is capable of melting 66 libs. 10 *oz.* 5 *gros* 24 *grs.* of ice.

One pound of charcoal during combustion melts only 96 *libs.* 8 *oz.* of ice, whilst it absorbs 2 *libs.* 9 *oz.* 1 *gros* 10 *grs.* of oxygen. By the experiment with phosphorus, this quantity of oxygen gas ought to disengage a quantity of caloric sufficient to melt 171 *libs.* 6 *oz.* 5 *gros* of ice; consequently, during this experiment, a quantity of caloric, sufficient to melt 74 *libs.* 14 *oz.* 5 *gros* of ice disappears. Carbonic acid is not, like phosphoric acid, in a concrete state after combustion but in the state of gas, and requires to be united with caloric to enable it to[Pg 99] subsist in that state; the quantity of caloric missing in the last experiment is evidently employed for that purpose. When we divide that quantity by the weight of carbonic acid, formed by the combustion of one pound of charcoal, we find that the quantity of caloric necessary for changing one pound of carbonic acid from the concrete to the gasseous state, would be capable of melting 20 *libs.* 15 *oz.* 5 *gros* of ice.

We may make a similar calculation with the combustion of hydrogen gas and the consequent formation of water. During the combustion of one pound of hydrogen gas, 5 *libs.* 10 *oz.* 5 *gros* 24 *grs.* of oxygen gas are absorbed, and 295 *libs.* 9 *oz.* 3-1/2 *gros*of ice are melted. But 5 *libs.* 10 *oz.* 5 *gros* 24 *grs.* of oxygen gas, in changing from the aëriform to the solid state, loses, according to the experiment with phosphorus, enough of caloric to have melted 377 *libs.* 12 *oz.* 3 *gros* of ice. There is only disengaged, from the same quantity of oxygen, during its combustion with hydrogen gas, as much caloric as melts 295 *libs.* 2 *oz.* 3-

1/2 *gros*; wherefore there remains in the water at Zero (32°), formed, during this experiment, as much caloric as would melt 82*libs.* 9 *oz.* 7-1/2 *gros* of ice.

Hence, as 6 *libs.* 10 *oz.* 5 *gros* 24 *grs.* of water are formed from the combustion of one pound of hydrogen gas with 5 *libs.* 10 *oz.* 5 *gros* 24 *grs.* of oxygen, it follows that, in each[Pg 100] pound of water, at the temperature of Zero, (32°), there exists as much caloric as would melt 12 *libs.* 5 *oz.* 2 *gros* 48 *grs.* of ice, without taking into account the quantity originally contained in the hydrogen gas, which we have been obliged to omit, for want of data to calculate its quantity. From this it appears that water, even in the state of ice, contains a considerable quantity of caloric, and that oxygen, in entering into that combination, retains likewise a good proportion.

From these experiments, we may assume the following results as sufficiently established.

Combustion of Phosphorus.

From the combustion of phosphorus, as related in the foregoing experiments, it appears, that one pound of phosphorus requires 1 *lib.* 8 *oz.* of oxygen gas for its combustion, and that 2 *libs.* 8 *oz.* of concrete phosphoric acid are produced.

The quantity of caloric disengaged by the combustion of one pound of phosphorus, expressed by the number of pounds of ice melted during that operation, is	100.00000.
The quantity disengaged from each pound of oxygen, during the combustion of phosphorus, expressed in the same manner, is	66.66667.
The quantity disengaged during the formation of one pound of phosphoric acid,	40.00000.
The quantity remaining in each pound of phosphoric acid,	0.00000(A).

[Note A: We here suppose the phosphoric acid not to contain any caloric, which is not strictly true; but, as I have before observed, the quantity it really contains is probably very small, and we have not given it a value, for want of a sufficient data to go upon.—A.][Pg 101]

Combustion of Charcoal.

In the combustion of one pound of charcoal, 2 *libs.* 9 *oz.* 1 *gros* 10 *grs.* of oxygen gas are absorbed, and 3 *libs.* 9 *oz.* 1 *gros* 10 *grs.* of carbonic acid gas are formed.

Caloric, disengaged during the combustion of one pound of charcoal,	96.50000(A).
Caloric disengaged during the combustion of charcoal, from each pound of oxygen gas absorbed,	37.52823.
Caloric disengaged during the formation of one pound of carbonic acid gas,	27.02024.
Caloric retained by each pound of oxygen after the combustion,	29.13844.
Caloric necessary for supporting one pound of carbonic acid in the state of gas,	20.97960.

[Note A: All these relative quantities of caloric are expressed by the number of pounds of ice, and decimal parts, melted during the several operations.—E.][Pg 102]

Combustion of Hydrogen Gas.

In the combustion of one pound of hydrogen gas, 5 *libs.* 10 *oz.* 5 *gros* 24 *grs.* of oxygen gas are absorbed, and 6 *libs.* 10 *oz.* 5 *gros* 24 *grs.* of water are formed.

Caloric from each *lib.* of hydrogen gas,	295.58950.
Caloric from each *lib.* of oxygen gas,	52.16280.
Caloric disengaged during the formation of each pound of water,	44.33840.
Caloric retained by each *lib.* of oxygen after combustion with hydrogen,	14.50386.
Caloric retained by each *lib.* of water at the temperature of Zero (32°),	12.32823.

Of the Formation of Nitric Acid.

When we combine nitrous gas with oxygen gas, so as to form nitric or nitrous acid a degree of heat is produced, which is much less considerable than what is evolved during the other combinations of oxygen; whence it follows that oxygen, when it becomes fixed in nitric acid, retains a great part of the heat which it possessed[Pg 103] in the state of gas. It is certainly possible to determine the quantity of caloric which is disengaged during the combination of these two gasses, and consequently to determine what quantity remains after the combination takes place. The first of these quantities might be ascertained, by making the combination of the two gasses in an apparatus surrounded by ice; but, as the quantity of caloric disengaged is very inconsiderable, it would be necessary to operate upon a large quantity of the two gasses in a very troublesome and complicated apparatus. By this consideration, Mr de la Place and I have hitherto been prevented from making the attempt. In the mean time, the place of such an experiment may be supplied by calculations, the results of which cannot be very far from truth.

Mr de la Place and I deflagrated a convenient quantity of nitre and charcoal in an ice apparatus, and found that twelve pounds of ice were melted by the deflagration of one pound of nitre. We shall see, in the sequel, that one pound of nitre is composed, as under, of

Potash	*oz.*		*gros*	5 1.84 *grs.*	451 5.84 *grs.*
Dry acid				2 1.16	470 0.16.

The above quantity of dry acid is composed of[Pg 104]

Oxygen	*oz.*		*gros*	6 6.34 *grs.*	373 8.34 *grs.*
Azote				2 5.82	961. 82.

By this we find that, during the above deflagration, 2 *gros* 1-1/3 *gr.* of charcoal have suffered combustion, alongst with 3738.34 *grs.* or 6 *oz.* 3 *gros* 66.34 *grs.* of oxygen. Hence, since 12 *libs.* of ice were melted during the combustion, it follows, that one pound of oxygen burnt in the same manner would have melted 29.58320 *libs.* of ice. To which the quantity of caloric, retained by a pound of oxygen after combining with charcoal to form carbonic acid gas, being added, which was already ascertained to be capable of melting 29.13844 *libs.* of ice, we have for the total quantity of caloric remaining in a pound of oxygen, when combined with nitrous gas in the nitric acid 58.72164; which is the number of pounds of ice the caloric remaining in the oxygen in that state is capable of melting.

We have before seen that, in the state of oxygen gas, it contained at least 66.66667; wherefore it follows that, in combining with azote to form nitric acid, it only loses 7.94502. Farther experiments upon this subject are necessary to ascertain how far the results of this calculation may agree with direct fact. This enormous quantity of caloric retained by oxygen in its combination into nitric acid, explains the[Pg 105] cause of the great disengagement of caloric during the deflagrations of nitre; or, more strictly speaking, upon all occasions of the decomposition of nitric acid.

Of the Combustion of Wax.

Having examined several cases of simple combustion, I mean now to give a few examples of a more complex nature. One pound of wax-taper being allowed to burn slowly in an ice apparatus, melted 133 *libs*. 2 *oz*. 5-1/3 *gros* of ice. According to my experiments in the Memoirs of the Academy for 1784, p. 606, one pound of wax-taper consists of 13 *oz*. 1 *gros* 23 *grs*. of charcoal, and 2 *oz*. 6 *gros* 49 *grs*. of hydrogen.

By the foregoing experiments, the above quantity of charcoal ought to melt   79.39390 *libs*. of ice;

and the hydrogen should melt   52.37605

In all   131.76995 *libs*.

Thus, we see the quantity of caloric disengaged from a burning taper, is pretty exactly conformable to what was obtained by burning separately a quantity of charcoal and hydrogen[Pg 106] equal to what enters into its composition. These experiments with the taper were several times repeated, so that I have reason to believe them accurate.

Combustion of Olive Oil.

We included a burning lamp, containing a determinate quantity of olive-oil, in the ordinary apparatus, and, when the experiment was finished, we ascertained exactly the quantities of oil consumed, and of ice melted; the result was, that, during the combustion of one pound of olive-oil, 148 *libs*. 14 *oz*. 1 *gros* of ice were melted. By my experiments in the Memoirs of the Academy for 1784, and of which the following Chapter contains an abstract, it appears that one pound of olive-oil consists of 12 *oz*. 5*gros* 5 *grs*. of charcoal, and 3 *oz*. 2 *gros* 67 *grs*. of hydrogen. By the foregoing experiments, that quantity of charcoal should melt 76.18723 *libs*. of ice, and the quantity of hydrogen in a pound of the oil should melt 62.15053 *libs*. The sum of these two gives 138.33776 *libs*. of ice, which the two constituent elements of the oil would have melted, had they separately suffered combustion, whereas the oil really melted 148.88330 *libs*. which gives an excess of 10.54554 in the result of the experiment[Pg 107] above the calculated result, from data furnished by former experiments.

This difference, which is by no means very considerable, may arise from errors which are unavoidable in experiments of this nature, or it may be owing to the composition of oil not being as yet exactly ascertained. It proves, however, that there is a great agreement between the results of our experiments, respecting the combination of caloric, and those which regard its disengagement.

The following desiderata still remain to be determined, viz. What quantity of caloric is retained by oxygen, after combining with metals, so as to convert them into oxyds; What quantity is contained by hydrogen, in its different states of existence; and to ascertain, with more precision than is hitherto attained, how much caloric is disengaged during the formation of water, as there still remain considerable doubts with respect to our present determination of this point, which can only be removed by farther experiments. We are at present occupied with this inquiry; and, when once these several points are well ascertained, which we hope they will soon be, we shall probably be under the necessity of making considerable corrections upon most of the results of the experiments and calculations in this Chapter. I did not, however, consider this as a sufficient reason for withholding[Pg 108] so much as is already known from such as may be inclined to labour upon the same subject. It is difficult, in our endeavours to discover the principles of a new science, to avoid beginning by guess-work; and it is rarely possible to arrive at perfection from the first setting out.

[Pg 109]
CHAP. X.
Of the Combination of Combustible Substances with each other.

As combustible substances in general have a great affinity for oxygen, they ought likewise to attract, or tend to combine with each other; *quae sunt eadem uni tertio, sunt eadem inter se*; and the axiom is found to be true. Almost all the metals, for instance, are capable of uniting with each other, and forming what are called *alloys*[22], in common language. Most of these, like all combinations, are susceptible of several degrees of saturation; the greater number of these alloys are more brittle than the pure metals of which they are composed, especially when the metals alloyed together are considerably different in their degrees of fusibility. To this difference in fusibility, part of the phenomena attendant upon *alloyage* are owing, particularly the property of iron, called by workmen [Pg 110] *hotshort*. This kind of iron must be considered as an alloy, or mixture of pure iron, which is almost infusible, with a small portion of some other metal which fuses in a much lower degree of heat. So long as this alloy remains cold, and both metals are in the solid state, the mixture is malleable; but, if heated to a sufficient degree to liquify the more fusible metal, the particles of the liquid metal, which are interposed between the particles of the metal remaining solid, must destroy their continuity, and occasion the alloy to become brittle. The alloys of mercury, with the other metals, have usually been called *amalgams*, and we see no inconvenience from continuing the use of that term.

Sulphur, phosphorus, and charcoal, readily unite with metals. Combinations of sulphur with metals are usually named *pyrites*. Their combinations with phosphorus and charcoal are either not yet named, or have received new names only of late; so that we have not scrupled to change them according to our principles. The combinations of metal and sulphur we call *sulphurets*, those with phosphorus *phosphurets*, and those formed with charcoal *carburets*. These denominations are extended to all the combinations into which the above three substances enter, without being previously oxygenated. [Pg 111] Thus, the combination of sulphur with potash, or fixed vegetable alkali, is called *sulphuret of potash*; that which it forms with ammoniac, or volatile alkali, is termed *sulphuret of ammoniac*.

Hydrogen is likewise capable of combining with many combustible substances. In the state of gas, it dissolves charcoal, sulphur, phosphorus, and several metals; we distinguish these combinations by the terms, *carbonated hydrogen gas, sulphurated hydrogen gas*, and *phosphorated hydrogen gas*. The sulphurated hydrogen gas was called *hepatic air* by former chemists, or *foetid air from sulphur*, by Mr Scheele. The virtues of several mineral waters, and the foetid smell of animal excrements, chiefly arise from the presence of this gas. The phosphorated hydrogen gas is remarkable for the property, discovered by Mr Gengembre, of taking fire spontaneously upon getting into contact with atmospheric air, or, what is better, with oxygen gas. This gas has a strong flavour, resembling that of putrid fish; and it is very probable that the phosphorescent quality of fish, in the state of putrefaction, arises from the escape of this species of gas. When hydrogen and charcoal are combined together, without the intervention of caloric, to bring the hydrogen into the state of gas, they form oil, which is either fixed or volatile, according to the proportions of hydrogen and [Pg 112] charcoal in its composition. The chief difference between fixed or fat oils drawn from vegetables by expression, and volatile or essential oils, is, that the former contains an excess of charcoal, which is separated when the oils are heated above the degree of boiling water; whereas the volatile oils, containing a just proportion of these two constituent ingredients, are not liable to be decomposed by that heat, but, uniting with caloric into the gasseous state, pass over in distillation unchanged.

In the Memoirs of the Academy for 1784, p. 593. I gave an account of my experiments upon the composition of oil and alkohol, by the union of hydrogen with charcoal, and of their combination with oxygen. By these experiments, it appears that fixed oils combine with oxygen during combustion, and are thereby converted into water and carbonic acid. By means of calculation applied to the products of these experiments, we find that fixed oil is composed of 21 parts, by weight, of hydrogen combined with 79 parts of charcoal. Perhaps the solid substances of an oily nature, such as wax, contain a proportion of oxygen, to which they owe their state of solidity. I am at present engaged in a series of experiments, which I hope will throw great light upon this subject.

It is worthy of being examined, whether hydrogen in its concrete state, uncombined with[Pg 113] caloric, be susceptible of combination with sulphur, phosphorus, and the metals. There is nothing that we know of, which, *a priori*, should render these combinations impossible; for combustible bodies being in general susceptible of combination with each other, there is no evident reason for hydrogen being an exception to the rule: However, no direct experiment as yet establishes either the possibility or impossibility of this union. Iron and zinc are the most likely, of all the metals, for entering into combination with hydrogen; but, as these have the property of decomposing water, and as it is very difficult to get entirely free from moisture in chemical experiments, it is hardly possible to determine whether the small portions of hydrogen gas, obtained in certain experiments with these metals, were previously combined with the metal in the state of solid hydrogen, or if they were produced by the decomposition of a minute quantity of water. The more care we take to prevent the presence of water in these experiments, the less is the quantity of hydrogen gas procured; and, when very accurate precautions are employed, even that quantity becomes hardly sensible.

However this inquiry may turn out respecting the power of combustible bodies, as sulphur, phosphorus, and metals, to absorb hydrogen, we are certain that they only absorb a very small[Pg 114] portion; and that this combination, instead of being essential to their constitution, can only be considered as a foreign substance, which contaminates their purity. It is the province of the advocates[23] for this system to prove, by decisive experiments, the real existence of this combined hydrogen, which they have hitherto only done by conjectures founded upon suppositions.

FOOTNOTES:

[22]This term *alloy*, which we have from the language of the arts, serves exceedingly well for distinguishing all the combinations or intimate unions of metals with each other, and is adopted in our new nomenclature for that purpose.—A.

[23]By these are meant the supporters of the phlogistic theory, who at present consider hydrogen, or the base of inflammable air, as the phlogiston of the celebrated Stahl.—E.

[Pg 115]

CHAP. XI.
Observations upon Oxyds and Acids with several Bases—and upon the Composition of Animal and Vegetable Substances.

We have, in Chap. V. and VIII. examined the products resulting from the combustion of the four simple combustible substances, sulphur, phosphorus, charcoal, and hydrogen: We have shown, in Chap. X that the simple combustible substances are capable of combining with each other into compound combustible substances, and have observed that oils in general, and particularly the fixed vegetable oils, belong to this class, being composed of hydrogen and charcoal. It remains, in this chapter, to treat of the oxygenation of these compound combustible substances, and to show that there exist acids and oxyds having double and triple bases. Nature furnishes us with numerous examples of this kind of combinations, by means of which, chiefly, she is enabled to produce a vast variety of compounds from a very limited number of elements, or simple substances.[Pg 116]

It was long ago well known, that, when muriatic and nitric acids were mixed together, a compound acid was formed, having properties quite distinct from those of either of the acids taken separately. This acid was called *aqua regia*, from its most celebrated property of dissolving gold, called *king of metals* by the alchymists. Mr Berthollet has distinctly proved that the peculiar properties of this acid arise from the combined action of its two acidifiable bases; and for this reason we have judged it necessary to distinguish it by an appropriate name: That of *nitro-muriatic* acid appears extremely applicable, from its expressing the nature of the two substances which enter into its composition.

This phenomenon of a double base in one acid, which had formerly been observed only in the nitro-muriatic acid, occurs continually in the vegetable kingdom, in which a simple acid, or one possessed of a single acidifiable base, is very rarely found. Almost all the acids procurable from this kingdom have bases composed of charcoal and hydrogen, or of charcoal, hydrogen, and phosphorus, combined with more or less oxygen. All these bases, whether double or triple, are likewise formed into oxyds, having less oxygen than is necessary to give them the properties of acids. The acids and oxyds from the animal kingdom are still more compound, as their bases generally consist of a combination[Pg 117] of charcoal, phosphorus, hydrogen, and azote.

As it is but of late that I have acquired any clear and distinct notions of these substances, I shall not, in this place, enlarge much upon the subject, which I mean to treat of very fully in some memoirs I am preparing to lay before the Academy. Most of my experiments are already performed; but, to be able to give exact reports of the resulting quantities, it is necessary that they be carefully repeated, and increased in number: Wherefore, I shall only give a short enumeration of the vegetable and animal acids and oxyds, and terminate this article by a few reflections upon the composition of vegetable and animal bodies.

Sugar, mucus, under which term we include the different kinds of gums, and starch, are vegetable oxyds, having hydrogen and charcoal combined, in different proportions, as their radicals or bases, and united with oxygen, so as to bring them to the state of oxyds. From the state of oxyds they are capable of being changed into acids by the addition of a fresh quantity of oxygen; and, according to the degrees of oxygenation, and the proportion of hydrogen and charcoal in their bases, they form the several kinds of vegetable acids.

It would be easy to apply the principles of our nomenclature to give names to these vegetable[Pg 118] acids and oxyds, by using the names of the two substances which compose their bases: They would thus become hydro-carbonous acids and oxyds: In this method we might indicate which of their elements existed in excess, without circumlocution, after the manner used by Mr Rouelle for naming vegetable extracts: He calls these extracto-resinous when the extractive matter prevails in their composition, and resino-extractive when they contain a larger proportion of resinous matter. Upon that plan, and by varying the terminations according to the formerly established rules of our nomenclature, we have the following denominations: Hydro-carbonous, hydro-carbonic; carbono-hydrous, and carbono-hydric oxyds. And for the acids: Hydro-carbonous, hydro carbonic, oxygenated hydro-carbonic; carbono-hydrous, carbono-hydric, and oxygenated carbono-hydric. It is probable that the above terms would suffice for indicating all the varieties in nature, and that, in proportion as the vegetable acids become well understood, they will naturally arrange themselves under these denominations. But, though we know the elements of which these are composed, we are as yet ignorant of the proportions of these ingredients, and are still far from being able to class them in the above methodical manner; wherefore, we have determined to retain[Pg 119] the ancient names provisionally. I am somewhat farther advanced in this inquiry than at the time of publishing our conjunct essay upon chemical nomenclature; yet it would be improper to draw decided consequences from experiments not yet sufficiently precise: Though I acknowledge that this part of chemistry still remains in some degree obscure, I must express my expectations of its being very soon elucidated.

I am still more forcibly necessitated to follow the same plan in naming the acids, which have three or four elements combined in their bases; of these we have a considerable number from the animal kingdom, and some even from vegetable substances. Azote, for instance, joined to hydrogen and charcoal, form the base or radical of the Prussic acid; we have reason to believe that the same happens with the base of the Gallic acid; and almost all the animal acids have their bases composed of azote, phosphorus, hydrogen, and charcoal. Were we to endeavour to express at once all these four component parts of the bases, our nomenclature would undoubtedly be methodical; it would have the property of being clear and determinate; but this assemblage of Greek and Latin substantives and adjectives, which are not yet universally admitted by chemists, would have the appearance of a[Pg 120] barbarous language, difficult both to pronounce and to be remembered. Besides, this part of chemistry being still far from that accuracy it must arrive to, the perfection of the science ought certainly to precede that of its language; and we must still, for some time, retain the old names for the animal oxyds and acids. We have only ventured to make a few slight modifications of these names, by changing the termination into *ous*, when we have reason to suppose the base to be in excess, and into *ic*, when we suspect the oxygen predominates.

The following are all the vegetable acids hitherto known:

1. Acetous acid.2. Acetic acid.3. Oxalic acid.4. Tartarous acid.5. Pyro-tartarous acid.6. Citric acid.7. Malic acid.8. Pyro-mucous acid.9. Pyro-lignous acid.10. Gallic acid.11. Benzoic acid.12. Camphoric acid.13. Succinic acid.

Though all these acids, as has been already said, are chiefly, and almost entirely, composed of hydrogen, charcoal, and oxygen, yet, properly speaking, they contain neither water carbonic acid nor oil, but only the elements necessary for forming these substances. The power of affinity reciprocally exerted by the hydrogen, charcoal, and oxygen, in these acids, is in a state[Pg 121] of equilibrium only capable of existing in the ordinary temperature of the atmosphere; for, when they are heated but a very little above the temperature of boiling water, this equilibrium is destroyed, part of the oxygen and hydrogen unite, and form water; part of the charcoal and hydrogen combine into oil; part of the charcoal and oxygen unite to form carbonic acid; and, lastly, there generally remains a small portion of charcoal, which, being in excess with respect to the other ingredients, is left free. I mean to explain this subject somewhat farther in the succeeding chapter.

The oxyds of the animal kingdom are hitherto less known than those from the vegetable kingdom, and their number is as yet not at all determined. The red part of the blood, lymph, and most of the secretions, are true oxyds, under which point of view it is very important to consider them. We are only acquainted with six animal acids, several of which, it is probable, approach very near each other in their nature, or, at least, differ only in a scarcely sensible degree. I do not include the phosphoric acid amongst these, because it is found in all the kingdoms of nature. They are,

1. Lactic acid.2. Saccholactic acid.3. Bombic acid.4. Formic acid.5. Sebacic acid.6. Prussic acid.[Pg 122]

The connection between the constituent elements of the animal oxyds and acids is not more permanent than in those from the vegetable kingdom, as a small increase of temperature is sufficient to overturn it. I hope to render this subject more distinct than has been done hitherto in the following chapter.

[Pg 123]

CHAP. XII.
Of the Decomposition of Vegetable and Animal Substances by the Action of Fire.

Before we can thoroughly comprehend what takes place during the decomposition of vegetable substances by fire, we must take into consideration the nature of the elements which enter into their composition, and the different affinities which the particles of these elements exert upon each other, and the affinity which caloric possesses with them. The true constituent elements of vegetables are hydrogen, oxygen, and charcoal: These are common to all vegetables, and no vegetable can exist without them: Such other substances as exist in particular vegetables are only essential to the composition of those in which they are found, and do not belong to vegetables in general.

Of these elements, hydrogen and oxygen have a strong tendency to unite with caloric, and be converted into gas, whilst charcoal is a fixed element, having but little affinity with caloric. On the other hand, oxygen, which, in the usual temperature, tends nearly equally to unite with hydrogen and with charcoal, has a much stronger[Pg 124] affinity with charcoal when at the red heat[24], and then unites with it to form carbonic acid.

Although we are far from being able to appreciate all these powers of affinity, or to express their proportional energy by numbers, we are certain, that, however variable they may be when considered in relation to the quantity of caloric with which they are combined, they are all nearly in equilibrium in the usual temperature of the atmosphere; hence vegetables neither contain oil[25], water, nor carbonic acid, tho' they contain all the elements of these substances. The hydrogen is neither combined with the oxygen nor with the charcoal, and reciprocally; the particles of these three substances form a triple combination, which remains in equilibrium[Pg 125] whilst undisturbed by caloric but a very slight increase of temperature is sufficient to overturn this structure of combination.

If the increased temperature to which the vegetable is exposed does not exceed the heat of boiling water, one part of the hydrogen combines with the oxygen, and forms water, the rest of the hydrogen combines with a part of the charcoal, and forms volatile oil, whilst the remainder of the charcoal, being set free from its combination with the other elements, remains fixed in the bottom of the distilling vessel.

When, on the contrary, we employ a red heat, no water is formed, or, at least, any that may have been produced by the first application of the heat is decomposed, the oxygen having a greater affinity with the charcoal at this degree of heat, combines with it to form carbonic acid, and the hydrogen being left free from combination with the other elements, unites with caloric, and escapes in the state of hydrogen gas. In this high temperature, either no oil is formed, or, if any was produced during the lower temperature at the beginning of the experiment, it is decomposed by the action of the red heat. Thus the decomposition of vegetable matter, under a high temperature, is produced by the action of double and triple affinities; while the charcoal attracts the oxygen,[Pg 126] on purpose to form carbonic acid, the caloric attracts the hydrogen, and converts it into hydrogen gas.

The distillation of every species of vegetable substance confirms the truth of this theory, if we can give that name to a simple relation of facts. When sugar is submitted to distillation, so long as we only employ a heat but a little below that of boiling water, it only loses its water of cristallization, it still remains sugar, and retains all its properties; but, immediately upon raising the heat only a little above that degree, it becomes blackened, a part of the charcoal separates from the combination, water slightly acidulated passes over accompanied by a little oil, and the charcoal which remains in the retort is nearly a third part of the original weight of the sugar.

The operation of affinities which take place during the decomposition, by fire, of vegetables which contain azote, such as the cruciferous plants, and of those containing phosphorus, is more complicated; but, as these substances only enter into the composition of vegetables in very small quantities, they only, apparently, produce slight changes upon the products of distillation; the phosphorus seems to combine with the charcoal, and, acquiring fixity from that union, remains behind in the retort, while the[Pg 127] azote, combining with a part of the hydrogen, forms ammoniac, or volatile alkali.

Animal substances, being composed nearly of the same elements with cruciferous plants, give the same products in distillation, with this difference, that, as they contain a greater quantity of hydrogen and azote, they produce more oil and more ammoniac. I shall only produce one fact as a proof of the exactness with which this theory explains all the phenomena which occur during the distillation of animal substances, which is the rectification and total decomposition of volatile animal oil, commonly known by the name of Dippel's oil. When these oils are procured by a first distillation in a naked fire they are brown, from containing a little charcoal almost in a free state; but they become quite colourless by rectification. Even in this state the charcoal in their composition has so slight a connection with the other elements as to separate by mere exposure to the air. If we put a quantity of this animal oil, well rectified, and consequently clear, limpid, and transparent, into a bell-glass filled with oxygen gas over mercury, in a short time the gas is much diminished, being absorbed by the oil, the oxygen combining with the hydrogen of the oil forms water, which sinks to the bottom, at the same time the charcoal which was combined with the hydrogen being set free, manifests itself[Pg 128] by rendering the oil black. Hence the only way of preserving these oils colourless and transparent, is by keeping them in bottles perfectly full and accurately corked, to hinder the contact of air, which always discolours them.

Successive rectifications of this oil furnish another phenomenon confirming our theory. In each distillation a small quantity of charcoal remains in the retort, and a little water is formed by the union of the oxygen contained in the air of the distilling vessels with the hydrogen of the oil. As this takes place in each successive distillation, if we make use of large vessels and a considerable degree of heat, we at last decompose the whole of the oil, and change it entirely into water and charcoal. When we use small vessels, and especially when we employ a slow fire, or degree of heat little above that of boiling water, the total decomposition of these oils, by repeated distillation, is greatly more tedious, and more difficultly accomplished. I shall give a particular detail to the Academy, in a separate memoir, of all my experiments upon the decomposition of oil; but what I have related above may suffice to give just ideas of the composition of animal and vegetable substances, and of their decomposition by the action of fire.

FOOTNOTES:

[24] Though this term, red heat, does not indicate any absolutely determinate degree of temperature, I shall use it sometimes to express a temperature considerably above that of boiling water.—A.

[25] I must be understood here to speak of vegetables reduced to a perfectly dry state; and, with respect to oil, I do not mean that which is procured by expression either in the cold, or in a temperature not exceeding that of boiling water; I only allude to the empyreumatic oil procured by distillation with a naked fire, in a heat superior to the temperature of boiling water; which is the only oil declared to be produced by the operation of fire. What I have published upon this subject in the Memoirs of the Academy for 1786 may be consulted.—A.

[Pg 129]

CHAP. XIII.
Of the Decomposition of Vegetable Oxyds by the Vinous Fermentation.

The manner in which wine, cyder, mead, and all the liquors formed by the spiritous fermentation, are produced, is well known to every one. The juice of grapes or of apples being expressed, and the latter being diluted with water, they are put into large vats, which are kept in a temperature of at least 10° (54.5°) of the thermometer. A rapid intestine motion, or fermentation, very soon takes place, numerous globules of gas form in the liquid and burst at the surface; when the fermentation is at its height, the quantity of gas disengaged is so great as to make the liquor appear as if boiling violently over a fire. When this gas is carefully gathered, it is found to be carbonic acid perfectly pure, and free from admixture with any other species of air or gas whatever.

When the fermentation is completed, the juice of grapes is changed from being sweet, and full of sugar, into a vinous liquor which no longer contains any sugar, and from which we procure, by distillation, an inflammable liquor,[Pg 130] known in commerce under the name of Spirit of Wine. As this liquor is produced by the fermentation of any saccharine matter whatever diluted with water, it must have been contrary to the principles of our nomenclature to call it spirit of wine rather than spirit of cyder, or of fermented sugar; wherefore, we have adopted a more general term, and the Arabic word *alkohol* seems extremely proper for the purpose.

This operation is one of the most extraordinary in chemistry: We must examine whence proceed the disengaged carbonic acid and the inflammable liquor produced, and in what manner a sweet vegetable oxyd becomes thus converted into two such opposite substances, whereof one is combustible, and the other eminently the contrary. To solve these two questions, it is necessary to be previously acquainted with the analysis of the fermentable substance, and of the products of the fermentation. We may lay it down as an incontestible axiom, that, in all the operations of art and nature, nothing is created; an equal quantity of matter exists both before and after the experiment; the quality and quantity of the elements remain precisely the same; and nothing takes place beyond changes and modifications in the combination of these elements. Upon this principle the whole art of performing chemical experiments[Pg 131] depends: We must always suppose an exact equality between the elements of the body examined and those of the products of its analysis.

Hence, since from must of grapes we procure alkohol and carbonic acid, I have an undoubted right to suppose that must consists of carbonic acid and alkohol. From these premises, we have two methods of ascertaining what passes during vinous fermentation, by determining the nature of, and the elements which compose, the fermentable substances, or by accurately examining the produces resulting from fermentation; and it is evident that the knowledge of either of these must lead to accurate conclusions concerning the nature and composition of the other. From these considerations, it became necessary accurately to determine the constituent elements of the fermentable substances; and, for this purpose, I did not make use of the compound juices of fruits, the rigorous analysis of which is perhaps impossible, but made choice of sugar, which is easily analysed, and the nature of which I have already explained. This substance is a true vegetable oxyd with two bases, composed of hydrogen and charcoal brought to the state of an oxyd, by a certain proportion of oxygen; and these three elements are combined in such a way, that a very slight force is sufficient to destroy the equilibrium of their connection. By[Pg 132] a long train of experiments, made in various ways, and often repeated, I ascertained that the proportion in which these ingredients exist in sugar, are nearly eight parts of hydrogen, 64 parts of oxygen, and 28 parts of charcoal, all by weight, forming 100 parts of sugar.

Sugar must be mixed with about four times its weight of water, to render it susceptible of fermentation; and even then the equilibrium of its elements would remain undisturbed, without the assistance of some substance, to give a commencement to the fermentation. This is accomplished by means of a little yeast from beer; and, when the fermentation is once excited, it continues of itself until completed. I shall, in another place, give an account of the effects of yeast, and other ferments, upon fermentable substances. I have usually employed 10 *libs.* of yeast, in the state of paste, for each 100 *libs.* of sugar, with as much water as is four times the weight of the sugar. I shall give the results of my experiments exactly as they were obtained, preserving even the fractions produced by calculation.[Pg 133]

TABLE I. *Materials of Fermentation.*

		ibs.	z	ros	rs.
Water		00			
Sugar		00			
Yeast in paste, 10 *libs.* composed of	{ Water			4	
	{ Dry yeast		2		8
	Total	10			

TABLE II. *Constituent Elements of the Materials of Fermentation.*

		l ibs.	*o* z	ros	*g* rs.
407 *libs*, 3 *oz.* 6 *gros* 44 *grs.* of water, composed of	{Hydrogen	61	1	7	1.40
	{ Oxygen	346	2	4	4.60

100 *libs.* sugar, composed of	{Hydrogen	8	0	0
	{Oxygen	64	0	0
	{Charcoal	28	0	0
2 *libs.* 12 *oz.* 1 *gros* 28 *grs.* of dry yeast, composed of	{Hydrogen	0	4	9.30
	{Oxygen	1	10	28.76
	{Charcoal	0	12	59
	{Azote	0	0	2.94
	Total weight	510	0	0

[Pg 134]

TABLE III. *Recapitulation of these Elements.*

	ibs.	*oz.*	*gros*	*grs.*				
Oxygen:								
of the water	40	0	0	0}	*ibs.*	*oz.*	*gros*	*grs.*
of the water in the yeast	2	3	4.60}		44	11	2	.36
of the sugar	4	0	0	0}				
of the dry yeast		1 0	2	28.76}				
Hydrogen:								
of the water	0	0	0	0}				
of the water in the yeast		1	2	71.40}		9		.70

of the sugar		0	0	0 }			
of the dry yeast		4	5	9 .30 }			
Charcoal:							
of the sugar	8	0	0	0 }			
of the yeast		1 2	4	5 9.00 }	8	2	9.00
Azote of the yeast		-	-	}			.94
		—	—	—			
				In all	10		

Having thus accurately determined the nature and quantity of the constituent elements of the materials submitted to fermentation, we have next to examine the products resulting from that process. For this purpose, I placed the above 510 *libs.* of fermentable liquor in a proper[26] apparatus, by means of which I could accurately determine the quantity and quality of gas disengaged during the fermentation, and could even weigh every one of the products[Pg 135] separately, at any period of the process I judged proper. An hour or two after the substances are mixed together, especially if they are kept in a temperature of from 15° (65.75°) to 18° (72.5°) of the thermometer, the first marks of fermentation commence; the liquor turns thick and frothy, little globules of air are disengaged, which rise and burst at the surface; the quantity of these globules quickly increases, and there is a rapid and abundant production of very pure carbonic acid, accompanied with a scum, which is the yeast separating from the mixture. After some days, less or more according to the degree of heat, the intestine motion and disengagement of gas diminish; but these do not cease entirely, nor is the fermentation completed for a considerable time. During the process, 35 *libs.* 5 *oz.* 4*gros* 19 *grs.* of dry carbonic acid are disengaged, which carry alongst with them 13*libs.* 14 *oz.* 5 *gros* of water. There remains in the vessel 460 *libs.* 11 *oz.* 6 *gros* 53 *grs.* of vinous liquor, slightly acidulous. This is at first muddy, but clears of itself, and deposits a portion of yeast. When we separately analise all these substances, which is effected by very troublesome processes, we have the results as given in the following Tables. This process, with all the subordinate calculations and analyses, will be detailed at large in the Memoirs of the Academy.

[Pg 136]

TABLE IV. *Product of Fermentation.*

		ibs.	*z.*	*ros*	*rs.*
35 *libs.* 5 *oz.* 4 *gros* 19 *grs.* of carbonic acid, composed of	{Oxygen		5		4
	{Charcoal		4		7

408 *libs.* 15 *oz.* 5 *gros* 14 *grs.* of water, composed of	{Oxygen	47	0	9
	{Hydrogen	1		7
57 *libs.* 11 *oz.* 1 *gros* 58 *grs.* of dry alkohol, composed of	{Oxygen, combined with hydrogen	1		4
	{Hydrogen, combined with oxygen			
	{Hydrogen, combined with charcoal			
	{Charcoal, combined with hydrogen	6	1	3
2 *libs.* 8 *oz.* of dry acetous acid, composed of	{Hydrogen			
	{Oxygen		1	
	{Charcoal		0	
4 *libs.* 1 *oz.* 4 *gros* 3 *grs.* of residuum of sugar, composed of	{Hydrogen			7
	{Oxygen			7
	{Charcoal			3
	{Hydrogen			1
1 *lib.* 6 *oz.* 0 *gros* 5 *grs.* of dry yeast, composed of	{Oxygen		3	4
	{ Charcoal			0

{Azote   7

510 *libs.*   Total   10

[Pg 137]

TABLE V. *Recapitulation of the Products.*

		lbs.	*oz.*	*gros*	*grs.*
409 libs. 10 oz. 0 gros 54 grs. of oxygen contained in the	Water	47	3 0	1 0	0 9 5
	Carbonic acid	5	2	7	1 4 3
	Alkohol	1	3	6	1 4 6
	Acetous acid	1	1 1		4 0
	Residuum of sugar		2	9	7 7 2
	Yeast		0 3	1	1 4 1
28 libs. 12 oz. 5 gros 59 grs. of charcoal contained in the	Carbonic acid		9 4	1	2 7 5
	Alkohol	6	1 1	1	5 3 6
	Acetous acid		0 0	1	0 0
	Residuum of sugar		1	2	2 3 5
	Yeast		0	6	2 0 3
71 libs. 8 oz. 6 gros 66 grs. of hydrogen contained in the	Water	1	6	5	4 7 2
	Water of the alkohol		5	8	5 3
	Combined		4	0	5 0

49

	with the charcoal of the alko.				
	Acetous acid	0	2	4	0
	Residuum of sugar	0	5	1 7	6
	Yeast	0	2	2 1	4
	2 gros 37 grs. of azote in the yeast	0	0	2 7	3
		—	—	—	—
510 *libs.*	Total	10 5	0	0	0

In these results, I have been exact, even to grains; not that it is possible, in experiments of this nature, to carry our accuracy so far, but as the experiments were made only with a few pounds of sugar, and as, for the sake of comparison, I reduced the results of the actual experiments to the quintal or imaginary hundred[Pg 138] pounds, I thought it necessary to leave the fractional parts precisely as produced by calculation.

When we consider the results presented by these tables with attention, it is easy to discover exactly what occurs during fermentation. In the first place, out of the 100*libs.* of sugar employed, 4 *libs*. 1 *oz*. 4 *gros* 3 *grs*. remain, without having suffered decomposition; so that, in reality, we have only operated upon 95 *libs*. 14 *oz*. 3 *gros*69 *grs*. of sugar; that is to say, upon 61 *libs*. 6 *oz*. 45 *grs*. of oxygen, 7 *libs*. 10 *oz*. 6*gros* 6 *grs*. of hydrogen, and 26 *libs*. 13 *oz*. 5 *gros* 19 *grs*. of charcoal. By comparing these quantities, we find that they are fully sufficient for forming the whole of the alkohol, carbonic acid and acetous acid produced by the fermentation. It is not, therefore, necessary to suppose that any water has been decomposed during the experiment, unless it be pretended that the oxygen and hydrogen exist in the sugar in that state. On the contrary, I have already made it evident that hydrogen, oxygen and charcoal, the three constituent elements of vegetables, remain in a state of equilibrium or mutual union with each other which subsists so long as this union remains undisturbed by increased temperature, or by some new compound attraction; and that then[Pg 139] only these elements combine, two and two together, to form water and carbonic acid.

The effects of the vinous fermentation upon sugar is thus reduced to the mere separation of its elements into two portions; one part is oxygenated at the expence of the other, so as to form carbonic acid, whilst the other part, being deoxygenated in favour of the former, is converted into the combustible substance alkohol; therefore, if it were possible to reunite alkohol and carbonic acid together, we ought to form sugar. It is evident that the charcoal and hydrogen in the alkohol do not exist in the state of oil, they are combined with a portion of oxygen, which renders them miscible with water; wherefore these three substances, oxygen, hydrogen, and charcoal, exist here likewise in a species of equilibrium or reciprocal combination; and in fact, when they are made to pass through a red hot tube of glass or porcelain, this union or equilibrium is destroyed, the elements become combined, two and two, and water and carbonic acid are formed.

I had formally advanced, in my first Memoirs upon the formation of water, that it was decomposed in a great number of chemical experiments, and particularly during the vinous fermentation. I then supposed that water existed ready formed in sugar, though I am now convinced that sugar only contains the elements[Pg 140] proper for composing it. It may be readily conceived, that it must have cost me a good deal to abandon my first notions, but by several years reflection, and after a great number of experiments and observations upon vegetable substances, I have fixed my ideas as above.

I shall finish what I have to say upon vinous fermentation, by observing, that it furnishes us with the means of analysing sugar and every vegetable fermentable matter. We may consider the substances submitted to fermentation, and the products resulting from that operation, as forming an algebraic equation; and, by successively supposing each of the elements in this equation unknown, we can calculate their values in succession, and thus verify our experiments by calculation, and our calculation by experiment reciprocally. I have often successfully employed this method for correcting the first results of my experiments, and to direct me in the proper road for repeating them to advantage. I have explained myself at large upon this subject, in a Memoir upon vinous fermentation already presented to the Academy, and which will speedily be published.

FOOTNOTES:

[26]The above apparatus is described in the Third Part.—A.

[Pg 141]

CHAP. XIV.
Of the Putrefactive Fermentation.

The phenomena of putrefaction are caused, like those of vinous fermentation, by the operation of very complicated affinities. The constituent elements of the bodies submitted to this process cease to continue in equilibrium in the threefold combination, and form themselves anew into binary combinations[27], or compounds, consisting of two elements only; but these are entirely different from the results produced by the vinous fermentation. Instead of one part of the hydrogen remaining united with part of the water and charcoal to form alkohol, as in the vinous fermentation, the whole of the hydrogen is dissipated, during putrefaction, in the form of hydrogen gas, whilst, at the same time, the oxygen and charcoal, uniting with caloric, escape in the form of carbonic acid gas; so that, when the whole process is finished, especially[Pg 142] if the materials have been mixed with a sufficient quantity of water, nothing remains but the earth of the vegetable mixed with a small portion of charcoal and iron. Thus putrefaction is nothing more than a complete analysis of vegetable substance, during which the whole of the constituent elements is disengaged in form of gas, except the earth, which remains in the state of mould[28].

Such is the result of putrefaction when the substances submitted to it contain only oxygen, hydrogen, charcoal and a little earth. But this case is rare, and these substances putrify imperfectly and with difficulty, and require a considerable time to complete their putrefaction. It is otherwise with substances containing azote, which indeed exists in all animal matters, and even in a considerable number of vegetable substances. This additional element is remarkably favourable to putrefaction; and for this reason animal matter is mixed with vegetable, when the putrefaction of these is wished to be hastened. The whole art of forming composts and dunghills, for the purposes of agriculture, consists in the proper application of this admixture.

The addition of azote to the materials of putrefaction not only accelerates the process,[Pg 143]that element likewise combines with part of the hydrogen, and forms a new substance called *volatile alkali* or *ammoniac*. The results obtained by analysing animal matters, by different processes, leave no room for doubt with regard to the constituent elements of ammoniac; whenever the azote has been previously separated from these substances, no ammoniac is produced; and in all cases they furnish ammoniac only in proportion to the azote they contain. This composition of ammoniac is likewise fully proved by Mr Berthollet, in the Memoirs of the Academy for 1785, p. 316. where he gives a variety of analytical processes by which ammoniac is decomposed, and its two elements, azote and hydrogen, procured separately.

I already mentioned in Chap. X. that almost all combustible bodies were capable of combining with each other; hydrogen gas possesses this quality in an eminent degree, it dissolves charcoal, sulphur, and phosphorus, producing the compounds named*carbonated* *hydrogen gas, sulphurated hydrogen gas,* and *phosphorated hydrogen gas.* The two latter of these gasses have a peculiarly disagreeable flavour; the sulphurated hydrogen gas has a strong resemblance to the smell of rotten eggs, and the phosphorated smells exactly like putrid fish. Ammoniac has likewise a peculiar odour, not less penetrating, or less disagreeable, than these other gasses. From[Pg 144] the mixture of these different flavours proceeds the fetor which accompanies the putrefaction of animal substances. Sometimes ammoniac predominates, which is easily perceived by its sharpness upon the eyes; sometimes, as in feculent matters, the sulphurated gas is most prevalent; and sometimes, as in putrid herrings, the phosphorated hydrogen gas is most abundant.

I long supposed that nothing could derange or interrupt the course of putrefaction; but Mr Fourcroy and Mr Thouret have observed some peculiar phenomena in dead bodies, buried at a certain depth, and preserved to a certain degree, from contact with air; having found the muscular flesh frequently converted into true animal fat. This must have arisen from the disengagement of the azote, naturally contained in the animal substance, by some unknown cause, leaving only the hydrogen and charcoal remaining, which are the elements proper for producing fat or oil. This observation upon the possibility of converting animal substances into fat may some time or other lead to discoveries of great importance to society. The faeces of animals, and other excrementitious matters, are chiefly composed of charcoal and hydrogen, and approach considerably to the nature of oil, of which they furnish a considerable quantity by distillation with a naked fire; but the intolerable foetor which accompanies all the products[Pg 145] of these substances prevents our expecting that, at least for a long time, they can be rendered useful in any other way than as manures.

I have only given conjectural approximations in this Chapter upon the composition of animal substances, which is hitherto but imperfectly understood. We know that they are composed of hydrogen, charcoal, azote, phosphorus, and sulphur, all of which, in a state of quintuple combination, are brought to the state of oxyd by a larger or smaller quantity of oxygen. We are, however, still unacquainted with the proportions in which these substances are combined, and must leave it to time to complete this part of chemical analysis, as it has already done with several others.

FOOTNOTES:

[27]Binary combinations are such as consist of two simple elements combined together. Ternary, and quaternary, consist of three and four elements.—E.

[28]In the Third Part will be given the description of an apparatus proper for being used in experiments of this kind.—A.

[Pg 146]

CHAP. XV.
Of the Acetous Fermentation.

The acetous fermentation is nothing more than the acidification or oxygenation of wine[29], produced in the open air by means of the absorption of oxygen. The resulting acid is the acetous acid, commonly called Vinegar, which is composed of hydrogen and charcoal united together in proportions not yet ascertained, and changed into the acid state by oxygen. As vinegar is an acid, we might conclude from analogy that it contains oxygen, but this is put beyond doubt by direct experiments: In the first place, we cannot change wine into vinegar without the contact of air containing oxygen; secondly, this process is accompanied by a diminution of the volume of the air in which it is carried on from the absorption of its oxygen; and, thirdly, wine may be changed into vinegar by any other means of oxygenation.

[Pg 147]

Independent of the proofs which these facts furnish of the acetous acid being produced by the oxygenation of wine, an experiment made by Mr Chaptal, Professor of Chemistry at Montpellier, gives us a distinct view of what takes place in this process. He impregnated water with about its own bulk of carbonic acid from fermenting beer, and placed this water in a cellar in vessels communicating with the air, and in a short time the whole was converted into acetous acid. The carbonic acid gas procured from beer vats in fermentation is not perfectly pure, but contains a small quantity of alkohol in solution, wherefore water impregnated with it contains all the materials necessary for forming the acetous acid. The alkohol furnishes hydrogen and one portion of charcoal, the carbonic acid furnishes oxygen and the rest of the charcoal, and the air of the atmosphere furnishes the rest of the oxygen necessary for changing the mixture into acetous acid. From this observation it follows, that nothing but hydrogen is wanting to convert carbonic acid into acetous acid; or more generally, that, by means of hydrogen, and according to the degree of oxygenation, carbonic acid may be changed into all the vegetable acids; and, on the contrary, that, by depriving any of the vegetable acids of their hydrogen, they may be converted into carbonic acid.[Pg 148]

Although the principal facts relating to the acetous acid are well known, yet numerical exactitude is still wanting, till furnished by more exact experiments than any hitherto performed; wherefore I shall not enlarge any farther upon the subject. It is sufficiently shown by what has been said, that the constitution of all the vegetable acids and oxyds is exactly conformable to the formation of vinegar; but farther experiments are necessary to teach us the proportion of the constituent elements in all these acids and oxyds. We may easily perceive, however, that this part of chemistry, like all the rest of its divisions, makes rapid progress towards perfection, and that it is already rendered greatly more simple than was formerly believed.

FOOTNOTES:

[29]The word Wine, in this chapter, is used to signify the liquor produced by the vinous fermentation, whatever vegetable substance may have been used for obtaining it.—E.

[Pg 149]

CHAP. XVI.
Of the Formation of Neutral Salts, and of their different Bases.

We have just seen that all the oxyds and acids from the animal and vegetable kingdoms are formed by means of a small number of simple elements, or at least of such as have not hitherto been susceptible of decomposition, by means of combination with oxygen; these are azote, sulphur, phosphorus, charcoal, hydrogen, and the muriatic radical[30]. We may justly admire the simplicity of the means employed by nature to multiply qualities and forms, whether by combining three or four acidifiable bases in different proportions, or by altering the dose of oxygen employed for oxydating or acidifying them. We shall find the means no less simple and diversified, and as abundantly productive of forms and qualities, in the order of bodies we are now about to treat of.

[Pg 150]

Acidifiable substances, by combining with oxygen, and their consequent conversion into acids, acquire great susceptibility of farther combination; they become capable of uniting with earthy and metallic bodies, by which means neutral salts are formed. Acids may therefore be considered as true *salifying* principles, and the substances with which they unite to form neutral salts may be called *salifiable* bases: The nature of the union which these two principles form with each other is meant as the subject of the present chapter.

This view of the acids prevents me from considering them as salts, though they are possessed of many of the principal properties of saline bodies, as solubility in water, &c. I have already observed that they are the result of a first order of combination, being composed of two simple elements, or at least of elements which act as if they were simple, and we may therefore rank them, to use the language of Stahl, in the order of *mixts*. The neutral salts, on the contrary, are of a secondary order of combination, being formed by the union of two *mixts* with each other, and may therefore be termed *compounds*. Hence I shall not arrange the alkalies[31] or earths in the class of salts, to which I[Pg 151] allot only such as are composed of an oxygenated substance united to a base.

I have already enlarged sufficiently upon the formation of acids in the preceding chapter, and shall not add any thing farther upon that subject; but having as yet given no account of the salifiable bases which are capable of uniting with them to form neutral salts, I mean, in this chapter, to give an account of the nature and origin of each of these bases. These are potash, soda, ammoniac, lime, magnesia, barytes, argill[32], and all the metallic bodies.

§ 1. *Of Potash.*

We have already shown, that, when a vegetable substance is submitted to the action of fire in distilling vessels, its component elements, oxygen, hydrogen, and charcoal, which formed a threefold combination in a state of equilibrium, unite, two and two, in obedience to affinities which act conformable to the degree of heat[Pg 152] employed. Thus, at the first application of the fire, whenever the heat produced exceeds the temperature of boiling water, part of the oxygen and hydrogen unite to form water; soon after the rest of the hydrogen, and part of the charcoal, combine into oil; and, lastly, when the fire is pushed to the red heat, the oil and water, which had been formed in the early part of the process, become again decomposed, the oxygen and charcoal unite to form carbonic acid, a large quantity of hydrogen gas is set free, and nothing but charcoal remains in the retort.

A great part of these phenomena occur during the combustion of vegetables in the open air; but, in this case, the presence of the air introduces three new substances, the oxygen and azote of the air and caloric, of which two at least produce considerable changes in the results of the operation. In proportion as the hydrogen of the vegetable, or that which results from the decomposition of the water, is forced out in the form of hydrogen gas by the progress of the fire, it is set on fire immediately upon getting in contact with the air, water is again formed, and the greater part of the caloric of the two gasses becoming free produces flame. When all the hydrogen gas is driven out, burnt, and again reduced to water, the remaining charcoal continues to burn, but without flame; it is[Pg 153] formed into carbonic acid, which carries off a portion of caloric sufficient to give it the gasseous form; the rest of the caloric, from the oxygen of the air, being set free, produces the heat and light observed during the combustion of charcoal. The whole vegetable is thus reduced into water and carbonic acid, and nothing remains but a small portion of gray earthy matter called ashes, being the only really fixed principles which enter into the constitution of vegetables.

The earth, or rather ashes, which seldom exceeds a twentieth part of the weight of the vegetable, contains a substance of a particular nature, known under the name of fixed vegetable alkali, or potash. To obtain it, water is poured upon the ashes, which dissolves the potash, and leaves the ashes which are insoluble; by afterwards evaporating the water, we obtain the potash in a white concrete form: It is very fixed even in a very high degree of heat. I do not mean here to describe the art of preparing potash, or the method of procuring it in a state of purity, but have entered upon the above detail that I might not use any word not previously explained.

The potash obtained by this process is always less or more saturated with carbonic acid, which is easily accounted for: As the potash does not form, or at least is not set free, but in proportion[Pg 154] as the charcoal of the vegetable is converted into carbonic acid by the addition of oxygen, either from the air or the water, it follows, that each particle of potash, at the instant of its formation, or at least of its liberation, is in contact with a particle of carbonic acid, and, as there is a considerable affinity between these two substances, they naturally combine together. Although the carbonic acid has less affinity with potash than any other acid, yet it is difficult to separate the last portions from it. The most usual method of accomplishing this is to dissolve the potash in water; to this solution add two or three times its weight of quick-lime, then filtrate the liquor and evaporate it in close vessels; the saline substance left by the evaporation is potash almost entirely deprived of carbonic acid. In this state it is soluble in an equal weight of water, and even attracts the moisture of the air with great avidity; by this property it furnishes us with an excellent means of rendering air or gas dry by exposing them to its action. In this state it is soluble in alkohol, though not when combined with carbonic acid; and Mr Berthollet employs this property as a method of procuring potash in the state of perfect purity.

All vegetables yield less or more of potash in consequence of combustion, but it is furnished in various degrees of purity by different vegetables; usually, indeed, from all of them it is[Pg 155] mixed with different salts from which it is easily separable. We can hardly entertain a doubt that the ashes, or earth which is left by vegetables in combustion, pre-existed in them before they were burnt, forming what may be called the skeleton, or osseous part of the vegetable. But it is quite otherwise with potash; this substance has never yet been procured from vegetables but by means of processes or intermedia capable of furnishing oxygen and azote, such as combustion, or by means of nitric acid; so that it is not yet demonstrated that potash may not be a produce from these operations. I have begun a series of experiments upon this object, and hope soon to be able to give an account of their results.

§ 2. *Of Soda.*

Soda, like potash, is an alkali procured by lixiviation from the ashes of burnt plants, but only from those which grow upon the sea-side, and especially from the herb *kali*, whence is derived the name *alkali*, given to this substance by the Arabians. It has some properties in common with potash, and others which are entirely different: In general, these two substances have peculiar characters in their saline combinations which are proper to each, and consequently distinguish them from each other; thus soda,[Pg 156] which, as obtained from marine plants, is usually entirely saturated with carbonic acid, does not attract the humidity of the atmosphere like potash, but, on the contrary, desiccates, its cristals effloresce, and are converted into a white powder having all the properties of soda, which it really is, having only lost its water of cristallization.

Hitherto we are not better acquainted with the constituent elements of soda than with those of potash, being equally uncertain whether it previously existed ready formed in the vegetable or is a combination of elements effected by combustion. Analogy leads us to suspect that azote is a constituent element of all the alkalies, as is the case with ammoniac; but we have only slight presumptions, unconfirmed by any decisive experiments, respecting the composition of potash and soda.

§ 3. *Of Ammoniac.*

We have, however, very accurate knowledge of the composition of ammoniac, or volatile alkali, as it is called by the old chemists. Mr Berthollet, in the Memoirs of the Academy for 1784, p. 316. has proved by analysis, that 1000 parts of this substance consist of about 807 parts of azote combined with 193 parts of hydrogen.[Pg 157]

Ammoniac is chiefly procurable from animal substances by distillation, during which process the azote and hydrogen necessary to its formation unite in proper proportions; it is not, however, procured pure by this process, being mixed with oil and water, and mostly saturated with carbonic acid. To separate these substances it is first combined with an acid, the muriatic for instance, and then disengaged from that combination by the addition of lime or potash. When ammoniac is thus produced in its greatest degree of purity it can only exist under the gasseous form, at least in the usual temperature of the atmosphere, it has an excessively penetrating smell, is absorbed in large quantities by water, especially if cold and assisted by compression. Water thus saturated with ammoniac has usually been termed volatile alkaline fluor; we shall call it either simply ammoniac, or liquid ammoniac, and ammoniacal gas when it exists in the aëriform state.

§ 4. *Of Lime, Magnesia, Barytes, and Argill.*

The composition of these four earths is totally unknown, and, until by new discoveries their constituent elements are ascertained, we are certainly authorised to consider them as simple bodies. Art has no share in the production of these earths, as they are all procured ready formed[Pg 158] from nature; but, as they have all, especially the three first, great tendency to combination, they are never found pure. Lime is usually saturated with carbonic acid in the state of chalk, calcarious spars, most of the marbles, &c.; sometimes with sulphuric acid, as in gypsum and plaster stones; at other times with fluoric acid forming vitreous or fluor spars; and, lastly, it is found in the waters of the sea, and of saline springs, combined with muriatic acid. Of all the salifiable bases it is the most universally spread through nature.

Magnesia is found in mineral waters, for the most part combined with sulphuric acid; it is likewise abundant in sea-water, united with muriatic acid; and it exists in a great number of stones of different kinds.

Barytes is much less common than the three preceding earths; it is found in the mineral kingdom, combined with sulphuric acid, forming heavy spars, and sometimes, though rarely, united to carbonic acid.

Argill, or the base of alum, having less tendency to combination than the other earths, is often found in the state of argill, uncombined with any acid. It is chiefly procurable from clays, of which, properly speaking, it is the base, or chief ingredient.[Pg 159]

§ 5. *Of Metallic Bodies.*

The metals, except gold, and sometimes silver, are rarely found in the mineral kingdom in their metallic state, being usually less or more saturated with oxygen, or combined with sulphur, arsenic, sulphuric acid, muriatic acid, carbonic acid, or phosphoric acid. Metallurgy, or the docimastic art, teaches the means of separating them from these foreign matters; and for this purpose we refer to such chemical books as treat upon these operations.

We are probably only acquainted as yet with a part of the metallic substances existing in nature, as all those which have a stronger affinity to oxygen, than charcoal possesses, are incapable of being reduced to the metallic state, and, consequently, being only presented to our observation under the form of oxyds, are confounded with earths. It is extremely probable that barytes, which we have just now arranged with earths, is in this situation; for in many experiments it exhibits properties nearly approaching to those of metallic bodies. It is even possible that all the substances we call earths may be only metallic oxyds, irreducible by any hitherto known process.

Those metallic bodies we are at present acquainted with, and which we can reduce to the[Pg 160] metallic or reguline state, are the following seventeen:

1. Arsenic.2. Molybdena.3. Tungstein.4. Manganese.5. Nickel.6. Cobalt.7. Bismuth.8. Antimony.9. Zinc.10. Iron.11. Tin.12. Lead.13. Copper.14. Mercury.15. Silver.16. Platina.17. Gold.

I only mean to consider these as salifiable bases, without entering at all upon the consideration of their properties in the arts, and for the uses of society. In these points of view each metal would require a complete treatise, which would lead me far beyond the bounds I have prescribed for this work.

FOOTNOTES:

[30]I have not ventured to omit this element, as here enumerated with the other principles of animal and vegetable substances, though it is not at all taken notice of in the preceding chapters as entering into the composition of these bodies.—E.

[31]Perhaps my thus rejecting the alkalies from the class of salts may be considered as a capital defect in the method I have adopted, and I am ready to admit the charge; but this inconvenience is compensated by so many advantages, that I could not think it of sufficient consequence to make me alter my plan.—A.

[32]Called Alumine by Mr Lavoisier; but as Argill has been in a manner naturalized to the language for this substance by Mr Kirwan, I have ventured to use it in preference.—E.

[Pg 161]

CHAP. XVII.
Continuation of the Observations upon Salifiable Bases, and the Formation of Neutral Salts.

It is necessary to remark, that earths and alkalies unite with acids to form neutral salts without the intervention of any medium, whereas metallic substances are incapable of forming this combination without being previously less or more oxygenated; strictly speaking, therefore, metals are not soluble in acids, but only metallic oxyds. Hence, when we put a metal into an acid for solution, it is necessary, in the first place, that it become oxygenated, either by attracting oxygen from the acid or from the water; or, in other words, that a metal cannot be dissolved in an acid unless the oxygen, either of the acid, or of the water mixed with it, has a stronger affinity to the metal than to the hydrogen or the acidifiable base; or, what amounts to the same thing, that no metallic solution can take place without a previous decomposition of the water, or the acid in which it is made. The explanation of the principal phenomena of metallic solution depends entirely[Pg 162] upon this simple observation, which was overlooked even by the illustrious Bergman.

The first and most striking of these is the effervescence, or, to speak less equivocally, the disengagement of gas which takes place during the solution; in the solutions made in nitric acid this effervescence is produced by the disengagement of nitrous gas; in solutions with sulphuric acid it is either sulphurous acid gas or hydrogen gas, according as the oxydation of the metal happens to be made at the expence of the sulphuric acid or of the water. As both nitric acid and water are composed of elements which, when separate, can only exist in the gasseous form, at least in the common temperature of the atmosphere, it is evident that, whenever either of these is deprived of its oxygen, the remaining element must instantly expand and assume the state of gas; the effervescence is occasioned by this sudden conversion from the liquid to the gasseous state. The same decomposition, and consequent formation of gas, takes place when solutions of metals are made in sulphuric acid: In general, especially by the humid way, metals do not attract all the oxygen it contains; they therefore reduce it, not into sulphur, but into sulphurous acid, and as this acid can only exist as gas in the usual temperature, it is disengaged, and occasions effervescence.[Pg 163]

The second phenomenon is, that, when the metals have been previously oxydated, they all dissolve in acids without effervescence: This is easily explained; because, not having now any occasion for combining with oxygen, they neither decompose the acid nor the water by which, in the former case, the effervescence is occasioned.

A third phenomenon, which requires particular consideration, is, that none of the metals produce effervescence by solution in oxygenated muriatic acid. During this process the metal, in the first place, carries off the excess of oxygen from the oxygenated muriatic acid, by which it becomes oxydated, and reduces the acid to the state of ordinary muriatic acid. In this case there is no production of gas, not that the muriatic acid does not tend to exist in the gasseous state in the common temperature, which it does equally with the acids formerly mentioned, but because this acid, which otherwise would expand into gas, finds more water combined with the oxygenated muriatic acid than is necessary to retain it in the liquid form; hence it does not disengage like the sulphurous acid, but remains, and quietly dissolves and combines with the metallic oxyd previously formed from its superabundant oxygen.

The fourth phenomenon is, that metals are absolutely insoluble in such acids as have their[Pg 164] bases joined to oxygen by a stronger affinity than these metals are capable of exerting upon that acidifying principle. Hence silver, mercury, and lead, in their metallic states, are insoluble in muriatic acid, but, when previously oxydated, they become readily soluble without effervescence.

From these phenomena it appears that oxygen is the bond of union between metals and acids; and from this we are led to suppose that oxygen is contained in all substances which have a strong affinity with acids: Hence it is very probable the four eminently salifiable earths contain oxygen, and their capability of uniting with acids is produced by the intermediation of that element. What I have formerly noticed relative to these earths is considerably strengthened by the above considerations, viz. that they may very possibly be metallic oxyds, with which oxygen has a stronger affinity than with charcoal, and consequently not reducible by any known means.

All the acids hitherto known are enumerated in the following table, the first column of which contains the names of the acids according to the new nomenclature, and in the second column are placed the bases or radicals of these acids, with observations.[Pg 165]

Names of the Acids.	Names of the Bases, with Observations.
1. Sulphurous	}Sulphur.
2. Sulphuric	}
3. Phosphorous	}Phosphorus.

4. Phosphoric	}	
5. Muriatic	}	Muriatic radical or base, hitherto unknown.
6. Oxygenated muriatic	}	
7. Nitrous	}	
8. Nitric	}	Azote.
9. Oxygenated nitric	}	
10. Carbonic	Charcoal	
	}	The bases or radicals of all these acids
11. Acetous	}	seem to be formed by a combination
12. Acetic	}	of charcoal and hydrogen;
13. Oxalic	}	and the only difference seems to be
14. Tartarous	}	owing to the different proportions in
15. Pyro-tartarous	}	which these elements combine to form
16. Citric	}	their bases, and to the different doses
17. Malic	}	of oxygen in their acidification. A
18. Pyro-lignous	}	connected series of accurate experiments
19. Pyro-mucous	}	is still wanted upon this subject.
20. Gallic	}	Our knowledge of the bases of
21. Prussic	}	these acids is hitherto imperfect; we

22. Benzoic	} only know that they contain hydrogen	
23. Succinic	} and charcoal as principal elements,	
24. Camphoric	} and that the prussic acid contains	
25. Lactic	} azote.	
26. Saccholactic	}	

27. Bombic	} The base of these and all acids	
28. Formic	} procured from animal substances seems	
29. Sebacic	} to consist of charcoal, hydrogen,	
	} phosphorous, and azote.	

30. Boracic	} The bases of these two are hitherto
31. Fluoric	} entirely unknown.

32. Antimonic	Antimony.
33. Argentic	Silver.
34. Arseniac(A)	Arsenic.

[Pg 166]

35. Bismuthic	Bismuth.
36. Cobaltic	Cobalt.
37. Cupric	Copper.
38. Stannic	Tin.
39. Ferric	Iron.
40. Manganic	Manganese.
41. Mercuric(B)	Mercury.
42. Molybdic	Molybdena.
43. Nickolic	Nickel.
44. Auric	Gold.

59

45. Platinic	Platina.
46. Plumbic	Lead.
47. Tungstic	Tungstein.
48. Zincic	Zinc.

[Note A: This term swerves a little from the rule in making the name of this acid terminate in *ac* instead of *ic*. The base and acid are distinguished in French by *arsenic* and *arsenique*; but, having chosen the English termination *ic* to translate the French*ique*, I was obliged to use this small deviation.—E.]

[Note B: Mr Lavoisier has hydrargirique; but mercurius being used for the base or metal, the name of the acid, as above, is equally regular, and less harsh.—E.]

In this list, which contains 48 acids, I have enumerated 17 metallic acids hitherto very imperfectly known, but upon which Mr Berthollet is about to publish a very important work. It cannot be pretended that all the acids which exist in nature, or rather all the acidifiable bases, are yet discovered; but, on the other hand, there are considerable grounds for supposing that a more accurate investigation than has hitherto been attempted will diminish the number of the vegetable acids, by showing that several of these, at present considered as distinct acids, are only[Pg 167] modifications of others. All that can be done in the present state of our knowledge is, to give a view of chemistry as it really is, and to establish fundamental principles, by which such bodies as may be discovered in future may receive names, in conformity with one uniform system.

The known salifiable bases, or substances capable of being converted into neutral salts by union with acids, amount to 24; viz. 3 alkalies, 4 earths, and 17 metallic substances; so that, in the present state of chemical knowledge, the whole possible number of neutral salts amounts to 1152[33]. This number is upon the supposition that the metallic acids are capable of dissolving other metals, which is a new branch of chemistry not hitherto investigated, upon which depends all the metallic combinations named *vitreous*. There is reason to believe that many of these supposable saline combinations are not capable of being formed, which must greatly reduce the real number of neutral salts producible by nature and art. Even if we suppose the real number to amount only to five or six hundred species of possible neutral salts, it is evident that, were we to distinguish them, after[Pg 168] the manner of the ancients, either by the names of their first discoverers, or by terms derived from the substances from which they are procured, we should at last have such a confusion of arbitrary designations, as no memory could possibly retain. This method might be tolerable in the early ages of chemistry, or even till within these twenty years, when only about thirty species of salts were known; but, in the present times, when the number is augmenting daily, when every new acid gives us 24 or 48 new salts, according as it is capable of one or two degrees of oxygenation, a new method is certainly necessary. The method we have adopted, drawn from the nomenclature of the acids, is perfectly analogical, and, following nature in the simplicity of her operations, gives a natural and easy nomenclature applicable to every possible neutral salt.

In giving names to the different acids, we express the common property by the generical term *acid*, and distinguish each species by the name of its peculiar acidifiable base. Hence the acids formed by the oxygenation of sulphur, phosphorus, charcoal, &c. are called *sulphuric acid, phosphoric acid, carbonic acid,* &c. We thought it likewise proper to indicate the different degrees of saturation with oxygen, by different terminations of the same specific names.[Pg 169] Hence we distinguish between sulphurous and sulphuric, and between phosphorous and phosphoric acids, &c.

By applying these principles to the nomenclature of neutral salts, we give a common term to all the neutral salts arising from the combination of one acid, and distinguish the species by adding the name of the salifiable base. Thus, all the neutral salts having sulphuric acid in their composition are named *sulphats*; those formed by the phosphoric acid, *phosphats*, &c. The species being distinguished by the names of the salifiable bases gives us *sulphat of potash, sulphat of soda, sulphat of ammoniac,sulphat of lime, sulphat of iron*, &c. As we are acquainted with 24 salifiable bases, alkaline, earthy, and metallic, we have consequently 24 sulphats, as many phosphats, and so on through all the acids. Sulphur is, however, susceptible of two degrees of oxygenation, the first of which produces sulphurous, and the second, sulphuric acid; and, as the neutral salts produced by these two acids, have different properties, and are in fact different salts, it becomes necessary to distinguish these by peculiar terminations; we have therefore distinguished the neutral salts formed by the acids in the first or lesser degree of oxygenation, by changing the termination *at* into *ite*, as *sulphites, phosphites*[34], &c. Thus, oxygenated[Pg 170] or acidified sulphur, in its two degrees of oxygenation is capable of forming 48 neutral salts, 24 of which are sulphites, and as many sulphats; which is likewise the case with all the acids capable of two degrees of oxygenation[35].

It were both tiresome and unnecessary to follow these denominations through all the varieties of their possible application; it is enough to have given the method of naming the various salts, which, when once well understood, is easily applied to every possible combination. The name of the combustible and acidifiable body being once known, the names of the acid it is capable of forming, and of all the neutral combinations[Pg 171] the acid is susceptible of entering into, are most readily remembered. Such as require a more complete illustration of the methods in which the new nomenclature is applied will, in the Second Part of this book, find Tables which contain a full enumeration of all the neutral salts, and, in general, all the possible chemical combinations, so far as is consistent with the present state of our knowledge. To these I shall subjoin short explanations, containing the best and most simple means of procuring the different species of acids, and some account of the general properties of the neutral salts they produce.

I shall not deny, that, to render this work more complete, it would have been necessary to add particular observations upon each species of salt, its solubility in water and alkohol, the proportions of acid and of salifiable base in its composition, the quantity of its water of cristallization, the different degrees of saturation it is susceptible of, and, finally, the degree of force or affinity with which the acid adheres to the base. This immense work has been already begun by Messrs Bergman, Morveau, Kirwan, and other celebrated chemists, but is hitherto only in a moderate state of advancement, even the principles upon which it is founded are not perhaps sufficiently accurate.[Pg 172]

These numerous details would have swelled this elementary treatise to much too great a size; besides that, to have gathered the necessary materials, and to have completed all the series of experiments requisite, must have retarded the publication of this book for many years. This is a vast field for employing the zeal and abilities of young chemists, whom I would advise to endeavour rather to do well than to do much, and to ascertain, in the first place, the composition of the acids, before entering upon that of the neutral salts. Every edifice which is intended to resist the ravages of time should be built upon a sure foundation; and, in the present state of chemistry, to attempt discoveries by experiments, either not perfectly exact, or not sufficiently rigorous, will serve only to interrupt its progress, instead of contributing to its advancement.

FOOTNOTES:

[33] This number excludes all triple salts, or such as contain more than one salifiable base, all the salts whose bases are over or under saturated with acid, and those formed by the nitro-muriatic acid.—E.

[34] As all the specific names of the acids in the new nomenclature are adjectives, they would have applied severally to the various salifiable bases, without the invention of other terms, with perfect distinctness. Thus, *sulphurous potash*, and *sulphuric potash*, are equally distinct as *sulphite of potash*, and *sulphat of potash*; and have the advantage of being more easily retained in the memory, because more naturally arising from the acids themselves, than the arbitrary terminations adopted by Mr Lavoisier.—E.

[35] There is yet a third degree of oxygenation of acids, as the oxygenated muriatic and oxygenated nitric acids. The terms applicable to the neutral salts resulting from the union of these acids with salifiable bases is supplied by the Author in the Second Part of this Work. These are formed by prefixing the word *oxygenated* to the name of the salt produced by the

second degree of oxygenation. Thus, *oxygenated* muriat of potash,*oxygenated* nitrat of soda, &c.—E.

[Pg 173]

PART II.
Of the Combination of Acids with Salifiable Bases, and of the Formation of Neutral Salts.

INTRODUCTION.

If I had strictly followed the plan I at first laid down for the conduct of this work, I would have confined myself, in the Tables and accompanying observations which compose this Second Part, to short definitions of the several known acids, and abridged accounts of the processes by which they are obtainable, with a mere nomenclature or enumeration of the neutral salts which result from the combination of these acids with the various salifiable bases. But I afterwards found that the addition of similar Tables of all the simple substances which enter[Pg 174] into the composition of the acids and oxyds, together with the various possible combinations of these elements, would add greatly to the utility of this work, without being any great increase to its size. These additions, which are all contained in the twelve first sections of this Part, and the Tables annexed to these, form a kind of recapitulation of the first fifteen Chapters of the First Part: The rest of the Tables and Sections contain all the saline combinations.

It must be very apparent that, in this Part of the Work, I have borrowed greatly from what has been already published by Mr de Morveau in the First Volume of the *Encyclopedie par ordre des Matières*. I could hardly have discovered a better source of information, especially when the difficulty of consulting books in foreign languages is considered. I make this general acknowledgment on purpose to save the trouble of references to Mr de Morveau's work in the course of the following part of mine.[Pg 175]

TABLE OF SIMPLE SUBSTANCES.

Simple substances belonging to all the kingdoms of nature, which may be considered as the elements of bodies.

New Names.	Correspondent old Names.
Light	Light.
Caloric	{Heat. {Principle or element of heat. {Fire. Igneous fluid. {Matter of fire and of heat.
Oxygen	{Dephlogisticated air.

New Names.		Correspondent old names.
Azote gas.		{ Empyreal air.
		{ Vital air, or
		{ Base of vital air.
		{ Phlogisticated air or
		{ Mephitis, or its base.
Hydrogen		{ Inflammable air or gas,
		{ or the base of inflammable air.

Oxydable and Acidifiable simple Substance not Metallic.

New Names.	Correspondent old names.
Sulphur	}
Phosphorous	} The same names.
Charcoal	}
Muriatic radical	}
Fluoric radical	} Still unknown.
Boracic radical	}

Oxydable and Acidifiable simple Metallic Bodies

New Names.	Correspondent Old Names.
Antimony	} Antimony.
Arsenic	} Arsenic.
Bismuth	} Bismuth.

	New Names		Correspondent old Names
	Cobalt	}	Cobalt.
	Copper	}	Copper.
	Gold	}	Gold.
	Iron	}	Iron.
	Lead	} Regulus of	Lead.
	Manganese	}	Manganese.
	Mercury	}	Mercury.
	Molybdena	}	Molybdena.
	Nickel	}	Nickel.
	Platina	}	Platina.
	Silver	}	Silver.
	Tin	}	Tin.
	Tungstein	}	Tungstein.
	Zinc	}	Zinc.

[Pg 176]

Salifiable simple Earthy Substances.

New Names.	Correspondent old Names.
Lime	{Chalk, calcareous earth.
	{Quicklime.
Magnesia	{Magnesia, base of Epsom salt.

64

		{Calcined or caustic magnesia.
ytes	Bar	Barytes, or heavy earth.
ll	Argi	Clay, earth of alum.
x	Sile	Siliceous or vitrifiable earth.

SECT. I.—*Observations upon the Table of Simple Substances.*

The principle object of chemical experiments is to decompose natural bodies, so as separately to examine the different substances which enter into their composition. By consulting chemical systems, it will be found that this science of chemical analysis has made rapid progress in our own times. Formerly oil and salt were considered as elements of bodies, whereas later observation and experiment have shown that all salts, instead of being simple, are composed of an acid united to a base. The bounds of analysis have been greatly enlarged by modern discoveries[36]; the acids are shown to be composed of oxygen, as an acidifying principle common to all, united in each to a particular base. I have proved what Mr Haffenfratz had[Pg 177] before advanced, that these radicals of the acids are not all simple elements, many of them being, like the oily principle, composed of hydrogen and charcoal. Even the bases of neutral salts have been proved by Mr Berthollet to be compounds, as he has shown that ammoniac is composed of azote and hydrogen.

Thus, as chemistry advances towards perfection, by dividing and subdividing, it is impossible to say where it is to end; and these things we at present suppose simple may soon be found quite otherwise. All we dare venture to affirm of any substance is, that it must be considered as simple in the present state of our knowledge, and so far as chemical analysis has hitherto been able to show. We may even presume that the earths must soon cease to be considered as simple bodies; they are the only bodies of the salifiable class which have no tendency to unite with oxygen; and I am much inclined to believe that this proceeds from their being already saturated with that element. If so, they will fall to be considered as compounds consisting of simple substances, perhaps metallic, oxydated to a certain degree. This is only hazarded as a conjecture; and I trust the reader will take care not to confound what I have related as truths, fixed on the firm basis of observation and experiment, with mere hypothetical conjectures.[Pg 178]

The fixed alkalies, potash, and soda, are omitted in the foregoing Table, because they are evidently compound substances, though we are ignorant as yet what are the elements they are composed of.[Pg 179]

TABLE *of compound oxydable and acidifiable bases.*

Names of the radicals.

Oxydable or acidifiable	{ Nitro-muriatic radical or
base, from the mineral	{ base of the acid formerly
kingdom.	{ called aqua regia.

	{ Tartarous radical or base.	
	{ Malic.	}
	{ Citric.	}
	{ Pyro-lignous.	}
Oxydable or acidifiable	{ Pyro-mucous.	}
hydro-carbonous or	{ Pyro-tartarous.	}
carbono-hydrous radicals	{ Oxalic.	}
from the vegetable	{ Acetous.	}
kingdom.	{ Succinic.	} Radicals
	{ Benzoic.	}
	{ Camphoric.	}
	{ Gallic.	}
		}
Oxydable or acidifiable	{ Lactic.	}
radicals from the animal	{ Saccholactic.	}
kingdom, which	{ Formic.	}
mostly contain azote,	{ Bombic.	}
and frequently phosphorus.	{ Sebacic.	}
	{ Lithic.	}
	{ Prussic.	}

Note.—The radicals from the vegetable kingdom are converted by a first degree of oxygenation into vegetable oxyds, such as sugar, starch, and gum or mucus: Those of the animal kingdom by the same means form animal oxyds, as lymph, &c.—A.[Pg 180]

SECT. II.—*Observations upon the Table of Compound Radicals.*

The older chemists being unacquainted with the composition of acids, and not suspecting them to be formed by a peculiar radical or base for each, united to an acidifying principle or element common to all, could not consequently give any name to substances of which they had not the most distant idea. We had therefore to invent a new nomenclature for this subject, though we were at the same time sensible that this nomenclature must be susceptible of great modification when the nature of the compound radicals shall be better understood[37].

The compound oxydable and acidifiable radicals from the vegetable and animal kingdoms, enumerated in the foregoing table, are not hitherto reducible to systematic nomenclature, because their exact analysis is as yet unknown. We only know in general, by some experiments of my own, and some made by Mr Hassenfratz, that most of the vegetable acids, such as the tartarous, oxalic, citric, malic, acetous, pyro-tartarous, and pyromucous, have radicals composed of hydrogen and charcoal, combined in[Pg 181] such a way as to form single bases, and that these acids only differ from each other by the proportions in which these two substances enter into the composition of their bases, and by the degree of oxygenation which these bases have received. We know farther, chiefly from the experiments of Mr Berthollet, that the radicals from the animal kingdom, and even some of those from vegetables, are of a more compound nature, and, besides hydrogen and charcoal, that they often contain azote, and sometimes phosphorus; but we are not hitherto possessed of sufficiently accurate experiments for calculating the proportions of these several substances. We are therefore forced, in the manner of the older chemists, still to name these acids after the substances from which they are procured. There can be little doubt that these names will be laid aside when our knowledge of these substances becomes more accurate and extensive; the terms *hydro-carbonous, hydro-carbonic, carbono-hydrous*, and *carbono hydric*[38], will then become substituted for those we now employ, which will then only remain as testimonies of the imperfect state in which this part of chemistry was transmitted to us by our predecessors.

[Pg 182]

It is evident that the oils, being composed of hydrogen and charcoal combined, are true carbono-hydrous or hydro-carbonous radicals; and, indeed, by adding oxygen, they are convertible into vegetable oxyds and acids, according to their degrees of oxygenation. We cannot, however, affirm that oils enter in their entire state into the composition of vegetable oxyds and acids; it is possible that they previously lose a part either of their hydrogen or charcoal, and that the remaining ingredients no longer exist in the proportions necessary to constitute oils. We still require farther experiments to elucidate these points.

Properly speaking, we are only acquainted with one compound radical from the mineral kingdom, the nitro-muriatic, which is formed by the combination of azote with the muriatic radical. The other compound mineral acids have been much less attended to, from their producing less striking phenomena.

SECT. III.—*Observations upon the Combinations of Light and Caloric with different Substances.*

I have not constructed any table of the combinations of light and caloric with the various simple and compound substances, because our conceptions of the nature of these combinations are not hitherto sufficiently accurate. We[Pg 183] know, in general, that all bodies in nature are imbued, surrounded, and penetrated in every way with caloric, which fills up every interval left between their particles; that, in certain cases, caloric becomes fixed in bodies, so as to constitute a part even of their solid substance, though it more frequently acts upon them with a repulsive force, from which, or from its accumulation in bodies to a greater or lesser degree, the transformation of solids into fluids, and of fluids to aëriform elasticity, is entirely owing. We have employed the generic name *gas* to indicate this aëriform state of bodies produced by a sufficient accumulation of caloric; so that, when we wish to express the aëriform state of muriatic acid, carbonic acid, hydrogen, water, alkohol, &c. we do it by adding the word *gas* to their names; thus muriatic acid gas, carbonic acid gas, hydrogen gas, aqueous gas, alkoholic gas, &c.

The combinations of light, and its mode of acting upon different bodies, is still less known. By the experiments of Mr Berthollet, it appears to have great affinity with oxygen, is susceptible of combining with it, and contributes alongst with caloric to change it into the state of gas. Experiments upon vegetation give reason to believe that light combines with certain parts of vegetables, and that the green of their leaves, and the various colours of their flowers, is chiefly[Pg 184] owing to this combination. This much is certain, that plants which grow in darkness are perfectly white, languid, and unhealthy, and that to make them recover vigour, and to acquire their natural colours, the direct influence of light is absolutely necessary. Somewhat similar takes place even upon animals: Mankind degenerate to a certain degree when employed in sedentary manufactures, or from living in crowded houses, or in the narrow lanes of large cities; whereas they improve in their nature and constitution in most of the country labours which are carried on in the open air. Organization, sensation, spontaneous motion, and all the operations of life, only exist at the surface of the earth, and in places exposed to the influence of light. Without it nature itself would be lifeless and inanimate. By means of light, the benevolence of the Deity hath filled the surface of the earth with organization, sensation, and intelligence. The fable of Promotheus might perhaps be considered as giving a hint of this philosophical truth, which had even presented itself to the knowledge of the ancients. I have intentionally avoided any disquisitions relative to organized bodies in this work, for which reason the phenomena of respiration, sanguification, and animal heat, are not considered; but I hope, at some future time, to be able to elucidate these curious subjects.

[Trancriber's note: The following table has been split into four sections ease reading]

TABLE of the binary Combinations of Oxygen with simple Substances

	Names of the simple substances.	First degree of oxygenation.	
		New Names.	Ancient Names.
Combinations of oxygen with simple non-metallic substances.	Caloric	Oxygen gas	Vital or dephlogisticated air
	Hydrogen.	Water(A).	
	Azote	Nitrous oxyd, or base of nitrous gas	Nitrous gas or air
	Charcoal	Oxyd of charcoal, or carbonic oxyd	Unknown
	Sulphur	Oxyd of sulphur	Soft sulphur
	Phosphorus	Oxyd of phosphorus {Residuum from the combustion of phosphorus	
	Muriatic radical	Muriatic oxyd	Unknown
	Fluoric radical	Fluoric oxyd	Unknown
	Boracic radical	Boracic oxyd	Unknown

Combinations of oxygen with the simple metallic substances.	y	Antimony	Grey oxyd of antimony	Grey calx of antimony
		Silver	Oxyd of silver	Calx of silver
		Arsenic	Grey oxyd of arsenic	Grey calx of arsenic
		Bismuth	Grey oxyd of bismuth	Grey calx of bismuth
		Cobalt	Grey oxyd of cobalt	Grey calx of cobalt
		Copper	Brown oxyd of copper	Brown calx of copper
		Tin	Grey oxyd of tin	Grey calx of tin
		Iron	Black oxyd of iron	Martial ethiops
	se	Mangane	Black oxyd of manganese	Black calx of manganese
		Mercury	Black oxyd of mercury	Ethiops mineral(B)
	na	Molybde	Oxyd of molybdena	Calx of molybdena
		Nickel	Oxyd of nickel	Calx of nickel
		Gold	Yellow oxyd of gold	Yellow calx of gold
		Platina	Yellow oxyd of platina	Yellow calx of platina
		Lead	Grey oxyd of lead	Grey calx of lead
	n	Tungstei	Oxyd of Tungstein	Calx of Tungstein
		Zinc	Grey oxyd of zinc	Grey calx of zinc

	Names of the simple substances.	Second degree of oxygenation.	
		New Names.	Ancient Names.
Combinations of oxygen with simple non-metallic substances.	Caloric		
	Hydrogen.		
	Azote	Nitrous acid	Smoking nitrous acid
	Charcoal	Carbonous acid	Unknown
	Sulphur	Sulphurous acid	Sulphureous acid
	Phosphorus	Phosphorous acid	Volatile acid of phosphorus
	Muriatic radical	Muriatous acid	Unknown
	Fluoric radical	Fluorous acid	Unknown
	Boracic radical	Boracous acid	Unknown
Combinations of oxygen with the simple metallic substances.	Antimony	White oxyd of antimony	White calx of antimony, diaphoretic antimony
	Silver		
	Arsenic	White oxyd of arsenic	White calx of arsenic
	Bismuth	White oxyd of bismuth	White calx of bismuth
	Cobalt		
	Copper	Blue and green oxyds of	Blue and green calces of copper

		copper	
	Tin	White oxyd of tin	White calx of tin, or putty of tin
	Iron	Yellow and red oxyds of iron	Ochre and rust of iron
se	Mangane	White oxyd of manganese	White calx of manganese
	Mercury	Yellow and red oxyds of mercury	Turbith mineral, red precipitate, calcinated mercury, precipitate per se
na	Molybde		
	Nickel		
	Gold	Red oxyd of gold	Red calx of gold, purple precipitate of cassius
	Platina		
	Lead	Yellow and red oxyds of lead	Massicot and minium
n	Tungstei		
	Zinc	White oxyd of zinc	White calx of zinc, pompholix

	Names of the simple substances.	Third degree of oxygenation.	
Combinations of oxygen with simple non-metallic substances.		New Names.	Ancient Names.
	Caloric		
	Hydrogen.		
	Azote	Nitric	Pale, or not

		acid	smoking nitrous acid
	Charcoal	Carbonic acid	Fixed air
	Sulphur	Sulphuric acid	Vitriolic acid
	Phosphorus	Phosphoric acid	Phosphoric acid
	Muriatic radical	Muriatic acid	Marine acid
	Fluoric radical	Fluoric acid	Unknown till lately
	Boracic radical	Boracic acid	Homberg's sedative salt
Combinations of oxygen with the simple metallic substances.	Antimony	Antimonic acid	
	Silver	Argentic acid	
	Arsenic	Arseniac acid	Acid of arsenic
	Bismuth	Bismuthic acid	
	Cobalt	Cobaltic acid	
	Copper	Cupric acid	
	Tin	Stannic acid	
	Iron	Ferric acid	
	Manganese	Manganesic acid	
	Mercury	Mercuric acid	

Molybdena	Molybdic acid	Acid of molybdena
Nickel	Nickelic acid	
Gold	Auric acid	
Platina	Platinic acid	
Lead	Plumbic acid	
Tungstein	Tungstic acid	Acid of Tungstein
Zinc	Zincic acid	

	Names of the simple substances.		Fourth degree of oxygenation.	
		New Names.		Ancient Names.
Combinations of oxygen with simple non-metallic substances.	Caloric			
	Hydrogen.			
	Azote	Oxygenated nitric acid		Unknown acid
	Charcoal	Oxygenated carbonic acid		Unknown
	Sulphur	Oxygenated sulphuric acid		Unknown
	Phosphorus	Oxygenated phosphoric acid		Unknown
	Muriatic radical	Oxygenated muriatic acid		Dephlogisticated marine acid
	Fluoric radical			

73

	Boracic radical		
Combinations of oxygen with the simple metallic substances.	Antimony		
	Silver		
	Arsenic	Oxygenated arseniac acid	Unknown
	Bismuth		
	Cobalt		
	Copper		
	Tin		
	Iron		
	Manganese		
	Mercury		
	Molybdena	Oxygenated molybdic acid	Unknown
	Nickel		
	Gold		
	Platina		
	Lead		
	Tungstein	Oxygenated Tungstic acid	Unknown
	Zinc		

[Note A: Only one degree of oxygenation of hydrogen is hitherto known.—A.]

[Note B: Ethiops mineral is the sulphuret of mercury; this should have been called black precipitate of mercury.—E.][Pg 185]

SECT. IV.—*Observations upon the Combinations of Oxygen with the simple Substances.*

Oxygen forms almost a third of the mass of our atmosphere, and is consequently one of the most plentiful substances in nature. All the animals and vegetables live and grow in this immense magazine of oxygen gas, and from it we procure the greatest part of what we employ in experiments. So great is the reciprocal affinity between this element and other substances, that we cannot procure it disengaged from all combination. In the atmosphere it is united with caloric, in the state of oxygen gas, and this again is mixed with about two thirds of its weight of azotic gas.

Several conditions are requisite to enable a body to become oxygenated, or to permit oxygen to enter into combination with it. In the first place, it is necessary that the particles of the body to be oxygenated shall have less reciprocal attraction with each other than they have for the oxygen, which otherwise cannot possibly combine with them. Nature, in this case, may be assisted by art, as we have it in our power to diminish the attraction of the particles of bodies almost at will by heating them, or, in other words, by introducing caloric into the interstices[Pg 186] between their particles; and, as the attraction of these particles for each other is diminished in the inverse ratio of their distance, it is evident that there must be a certain point of distance of particles when the affinity they possess with each other becomes less than that they have for oxygen, and at which oxygenation must necessarily take place if oxygen be present.

We can readily conceive that the degree of heat at which this phenomenon begins must be different in different bodies. Hence, on purpose to oxygenate most bodies, especially the greater part of the simple substances, it is only necessary to expose them to the influence of the air of the atmosphere in a convenient degree of temperature. With respect to lead, mercury, and tin, this needs be but little higher than the medium temperature of the earth; but it requires a more considerable degree of heat to oxygenate iron, copper, &c. by the dry way, or when this operation is not assisted by moisture. Sometimes oxygenation takes place with great rapidity, and is accompanied by great sensible heat, light, and flame; such is the combustion of phosphorus in atmospheric air, and of iron in oxygen gas. That of sulphur is less rapid; and the oxygenation of lead, tin, and most of the metals, takes place vastly slower, and consequently the disengagement of caloric, and more especially of light, is hardly sensible.[Pg 187]

Some substances have so strong an affinity with oxygen, and combine with it in such low degrees of temperature, that we cannot procure them in their unoxygenated state; such is the muriatic acid, which has not hitherto been decomposed by art, perhaps even not by nature, and which consequently has only been found in the state of acid. It is probable that many other substances of the mineral kingdom are necessarily oxygenated in the common temperature of the atmosphere, and that being already saturated with oxygen, prevents their farther action upon that element.

There are other means of oxygenating simple substances besides exposure to air in a certain degree of temperature, such as by placing them in contact with metals combined with oxygen, and which have little affinity with that element. The red oxyd of mercury is one of the best substances for this purpose, especially with bodies which do not combine with that metal. In this oxyd the oxygen is united with very little force to the metal, and can be driven out by a degree of heat only sufficient to make glass red hot; wherefore such bodies as are capable of uniting with oxygen are readily oxygenated, by means of being mixed with red oxyd of mercury, and moderately heated. The same effect may be, to a certain degree, produced by means of the black oxyd of manganese, the red oxyd of lead,[Pg 188] the oxyds of silver, and by most of the metallic oxyds, if we only take care to choose such as have less affinity with oxygen than the bodies they are meant to oxygenate. All the metallic reductions and revivifications belong to this class of operations, being nothing more than oxygenations of charcoal, by means of the several metallic oxyds. The charcoal combines with the oxygen and with caloric, and escapes in form of carbonic acid gas, while the metal remains pure and revivified, or deprived of the oxygen which before combined with it in the form of oxyd.

All combustible substances may likewise be oxygenated by means of mixing them with nitrat of potash or of soda, or with oxygenated muriat of potash, and subjecting the mixture to a certain degree of heat; the oxygen, in this case, quits the nitrat or the muriat, and combines with the combustible body. This species of oxygenation requires to be performed with extreme caution, and only with very small quantities; because, as the oxygen enters into the composition of nitrats, and more especially of oxygenated muriats, combined with almost as much caloric as is necessary for converting it into oxygen gas, this immense quantity of caloric becomes suddenly free the instant of the combination of the oxygen with the combustible[Pg 189] body, and produces such violent explosions as are perfectly irresistible.

By the humid way we can oxygenate most combustible bodies, and convert most of the oxyds of the three kingdoms of nature into acids. For this purpose we chiefly employ the nitric acid, which has a very slight hold of oxygen, and quits it readily to a great number of bodies by the assistance of a gentle heat. The oxygenated muriatic acid may be used for several operations of this kind, but not in them all.

I give the name of *binary* to the combinations of oxygen with the simple substances, because in these only two elements are combined. When three substances are united in one combination I call it *ternary*, and *quaternary* when the combination consists of four substances united.[Pg 190]

TABLE *of the combinations of Oxygen with the compound radicals.*

Names of the radicals.	Names of the resulting acids.	
	New nomenclature.	*Old nomenclature.*
Nitro muriatic radical	Nitro muriatic acid	Aqua regia.
(A)		
Tartaric	Tartarous acid	Unknown till lately.
Malic	Malic acid	Ditto.
Citric	Citric acid	Acid of lemons.
Pyro-lignous	Pyro-lignous acid	Empyreumatic acid of wood.
Pyro-mucous	Pyro-mucous acid	Empyr. acid of sugar.
Pyro-tartarous	Pyro-tartarous acid	Empyr. acid of tartar.
Oxalic	Oxalic acid	Acid of sorel.
Acetic	{Acetous acid	Vinegar, or acid of vinegar.
	{Acetic acid	Radical vinegar.
Succinic	Succinic	Volatile salt of amber.

		acid	
Benzoic	acid	Benzotic acid	Flowers of benzoin.
Camphoric	acid	Camphoric acid	Unknown till lately.
Gallic		Gallic acid	The astringent principle of vegetables.

(B)

Lactic		Lactic acid	Acid of sour whey.
Saccholactic	c acid	Saccholacti	Unknown till lately.
Formic		Formic acid	Acid of ants.
Bombic	acid	Bombic	Unknown till lately.
Sebacic	acid	Sebacic	Ditto.
Lithic		Lithic acid	Urinary calculus.
Prussic		Prussic acid	Colouring matter of Prussian blue.

[Note A: These radicals by a first degree of oxygenation form vegetable oxyds, as sugar, starch, mucus, &c.—A.]

[Note B: These radicals by a first degree of oxygenation form the animal oxyds, as lymph, red part of the blood, animal secretions, &c.—A.][Pg 191]

SECT. V.—*Observations upon the Combinations of Oxygen with the Compound Radicals.*

I published a new theory of the nature and formation of acids in the Memoirs of the Academy for 1776, p. 671. and 1778, p. 535. in which I concluded, that the number of acids must be greatly larger than was till then supposed. Since that time, a new field of inquiry has been opened to chemists; and, instead of five or six acids which were then known, near thirty new acids have been discovered, by which means the number of known neutral salts have been increased in the same proportion. The nature of the acidifiable bases, or radicals of the acids, and the degrees of oxygenation they are susceptible of, still remain to be inquired into. I have already shown, that almost all the oxydable and acidifiable radicals from the mineral kingdom are simple, and that, on the contrary, there hardly exists any radical in the vegetable, and more especially in the animal kingdom, but is composed of at least two substances, hydrogen and charcoal, and that azote and phosphorus are frequently united to these, by which we have compound radicals of two, three, and four bases or simple elements united.[Pg 192]

From these observations, it appears that the vegetable and animal oxyds and acids may differ from each other in three several ways: 1st, According to the number of simple acidifiable elements of which their radicals are composed: 2dly, According to the proportions in which these are combined together: And, 3dly, According to their different degrees of oxygenation: Which circumstances are more than sufficient to explain the great variety which nature produces in these substances. It is not at all surprising, after this, that most of the vegetable acids are convertible into each other, nothing more being requisite than to change the proportions of the hydrogen and charcoal in their composition, and to oxygenate them in a greater or lesser degree. This has been done by Mr Crell in some very ingenious experiments, which have been verified and extended by Mr Hassenfratz. From these it appears, that charcoal and hydrogen, by a first oxygenation, produce tartarous acid, oxalic acid by a second degree, and acetous or acetic acid by a third, or higher oxygenation; only, that charcoal seems to exist in a rather smaller proportion in the acetous and acetic acids. The citric and malic acids differ little from the preceding acids.

Ought we then to conclude that the oils are the radicals of the vegetable and animal acids? I have already expressed my doubts upon this[Pg 193] subject: 1st, Although the oils appear to be formed of nothing but hydrogen and charcoal, we do not know if these are in the precise proportion necessary for constituting the radicals of the acids: 2dly, Since oxygen enters into the composition of these acids equally with hydrogen and charcoal, there is no more reason for supposing them to be composed of oil rather than of water or of carbonic acid. It is true that they contain the materials necessary for all these combinations, but then these do not take place in the common temperature of the atmosphere; all the three elements remain combined in a state of equilibrium, which is readily destroyed by a temperature only a little above that of boiling water[39].

[Pg 194]

TABLE *of the Binary Combinations of Azote with the Simple Substances.*

Simple Substances.	Results of the Combinations.	
	New Nomenclature.	Old Nomenclature.
Caloric	Azotic gas	Phlogisticated air, or Mephitis.
Hydrogen	Ammoniac	Volatile alkali.
Oxygen	{Nitrous oxyd	Base of Nitrous gas.
	{Nitrous acid	Smoaking nitrous acid.
	{Nitric acid	Pale nitrous acid.
	{Oxygenated nitric acid	Unknown.
Charcoal	{This combination is hitherto unknown; should it {ever be discovered, it will be called, according to {the principles of our nomenclature, Azuret of {Charcoal. Charcoal dissolves in azotic gas, and	

		{forms carbonated azotic gas.	
Phosphorus.		Azuret of phosphorus.	Still unknown.
	Sulphur	{Azuret of sulphur.	Still unknown. We know
		{that sulphur dissolves in azotic gas, forming	
		{sulphurated azotic gas.	
Compound radicals		{Azote combines with charcoal and hydrogen, and	
		{sometimes with phosphorus, in the compound	
		{oxydable and acidifiable bases, and is generally	
		{contained in the radicals of the animal acids.	
Metallic substances		{Such combinations are hitherto unknown; if ever	
		{discovered, they will form metallic azurets, as	
		{azuret of gold, of silver, &c.	
	Lime	{	
	Magnesia	{	
	Barytes	{Entirely unknown. If ever discovered, they will	
	Argill	{form azuret of lime, azuret of magnesia, &c.	
	Potash	{	

Soda　　　{

[Pg 195]

SECT. VI.—*Observations upon the Combinations of Azote with the Simple Substances.*

Azote is one of the most abundant elements; combined with caloric it forms azotic gas, or mephitis, which composes nearly two thirds of the atmosphere. This element is always in the state of gas in the ordinary pressure and temperature, and no degree of compression or of cold has been hitherto capable of reducing it either to a solid or liquid form. This is likewise one of the essential constituent elements of animal bodies, in which it is combined with charcoal and hydrogen, and sometimes with phosphorus; these are united together by a certain portion of oxygen, by which they are formed into oxyds or acids according to the degree of oxygenation. Hence the animal substances may be varied, in the same way with vegetables, in three different manners: 1st, According to the number of elements which enter into the composition of the base or radical: 2dly, According to the proportions of these elements: 3dly, According to the degree of oxygenation.

When combined with oxygen, azote forms the nitrous and nitric oxyds and acids; when with hydrogen, ammoniac is produced. Its combinations with the other simple elements[Pg 196] are very little known; to these we give the name of Azurets, preserving the termination in *uret* for all nonoxygenated compounds. It is extremely probable that all the alkaline substances may hereafter be found to belong to this genus of azurets.

The azotic gas may be procured from atmospheric air, by absorbing the oxygen gas which is mixed with it by means of a solution of sulphuret of potash, or sulphuret of lime. It requires twelve or fifteen days to complete this process, during which time the surface in contact must be frequently renewed by agitation, and by breaking the pellicle which forms on the top of the solution. It may likewise be procured by dissolving animal substances in dilute nitric acid very little heated. In this operation, the azote is disengaged in form of gas, which we receive under bell glasses filled with water in the pneumato-chemical apparatus. We may procure this gas by deflagrating nitre with charcoal, or any other combustible substance; when with charcoal, the azotic gas is mixed with carbonic acid gas, which may be absorbed by a solution of caustic alkali, or by lime water, after which the azotic gas remains pure. We can procure it in a fourth manner from combinations of ammoniac with metallic oxyds, as pointed out by Mr de Fourcroy: The hydrogen of the ammoniac combines with the oxygen of the[Pg 197] oxyd, and forms water, whilst the azote being left free escapes in form of gas.

The combinations of azote were but lately discovered: Mr Cavendish first observed it in nitrous gas and acid, and Mr Berthollet in ammoniac and the prussic acid. As no evidence of its decomposition has hitherto appeared, we are fully entitled to consider azote as a simple elementary substance.[Pg 198]

TABLE *of the Binary Combinations of Hydrogen with Simple Substances.*

Simple Substances.	Resulting Compounds.	
	New Nomenclature.	*Old Names.*
Caloric	Hydrogen gas　　air.	Inflammable
Azote	Ammoniac	Volatile Alkali.
Oxygen	Water	Water.
Sulphur	or } {Hydruret of sulphur,	

		{sulphuret of hydrogen }	Hitherto unknown (A).
rus	Phospho	{Hydruret of phosphorus, or } {phosphuret of hydrogen }	
	Charcoal	{Hydro-carbonous, or } {carbono-hydrous radicals(B)}	Not known till lately.
es, as	Metallic substanc iron, &c.	as} {Metallic hydrurets(C), {hydruret of iron, &c.} {	Hitherto unknown. }

[Note A: These combinations take place in the state of gas, and form, respectively, sulphurated and phosphorated oxygen gas—A.]

[Note B: This combination of hydrogen with charcoal includes the fixed and volatile oils, and forms the radicals of a considerable part of the vegetable and animal oxyds and acids. When it takes place in the state of gas it forms carbonated hydrogen gas.—A.]

[Note C: None of these combinations are known, and it is probable that they cannot exist, at least in the usual temperature of the atmosphere, owing to the great affinity of hydrogen for caloric.—A.][Pg 199]

SECT. VII.—*Observations upon Hydrogen, and its Combinations with Simple Substances.*

Hydrogen, as its name expresses, is one of the constituent elements of water, of which it forms fifteen hundredth parts by weight, combined with eighty-five hundredth parts of oxygen. This substance, the properties and even existence of which was unknown till lately, is very plentifully distributed in nature, and acts a very considerable part in the processes of the animal and vegetable kingdoms. As it possesses so great affinity with caloric as only to exist in the state of gas, it is consequently impossible to procure it in the concrete or liquid state, independent of combination.

To procure hydrogen, or rather hydrogen gas, we have only to subject water to the action of a substance with which oxygen has greater affinity than it has to hydrogen; by this means the hydrogen is set free, and, by uniting with caloric, assumes the form of hydrogen gas. Red hot iron is usually employed for this purpose: The iron, during the process, becomes oxydated, and is changed into a substance resembling the iron ore from the island of Elba. In this state of oxyd it is much less attractible by[Pg 200] the magnet, and dissolves in acids without effervescence.

Charcoal, in a red heat, has the same power of decomposing water, by attracting the oxygen from its combination with hydrogen. In this process carbonic acid gas is formed, and mixes with the hydrogen gas, but is easily separated by means of water or alkalies, which absorb the carbonic acid, and leave the hydrogen gas pure. We may likewise obtain hydrogen gas by dissolving iron or zinc in dilute sulphuric acid. These two metals decompose water very slowly, and with great difficulty, when alone, but do it with great ease and rapidity when assisted by sulphuric acid; the hydrogen unites with caloric during the process, and is disengaged in form of hydrogen gas, while the oxygen of the water unites with the metal in the form of oxyd, which is immediately dissolved in the acid, forming a sulphat of iron or of zinc.

Some very distinguished chemists consider hydrogen as the *phlogiston* of Stahl; and as that celebrated chemist admitted the existence of phlogiston in sulphur, charcoal, metals, &c. they are of course obliged to suppose that hydrogen exists in all these substances, though they cannot prove their supposition; even if they could, it would not avail much, since this disengagement of hydrogen is quite insufficient to explain the phenomena of calcination and combustion.[Pg 201] We must always recur to the examination of this question, "Are the heat and light, which are disengaged during the different species of combustion, furnished by the burning body, or by the oxygen which combines in all these operations?" And certainly the supposition of hydrogen being disengaged throws no light whatever upon this question. Besides, it belongs to those who make suppositions to prove them; and, doubtless, a doctrine which without any supposition explains the phenomena as well, and as naturally, as theirs does by supposition, has at least the advantage of greater simplicity[40].

[Pg 202]

TABLE *of the Binary Combinations of Sulphur with Simple Substances.*

Simple Substances.	Resulting Compounds.	
	New Nomenclature.	Old Nomenclature.
Caloric	Sulphuric gas	
Oxygen	{Oxyd of sulphur	Soft sulphur.
	{Sulphurous acid	Sulphureous acid.
	{Sulphuric acid	Vitriolic acid.
Hydrogen	Sulphuret of hydrogen}	
Azote	azote}	Unknown Combinations.
Phosphorus	phosphorus}	
Charcoal	charcoal}	
Antimony	antimony	Crude antimony.
Silver	silver	
Arsenic	arsenic	Orpiment, realgar.
Bismuth	bismuth	

Cobalt	cobalt	
Copper	copper	Copper pyrites.
Tin	tin	
Iron	iron	Iron pyrites.
Manganese	manganese	
Mercury	mercury	Ethiops mineral, cinnabar.
Molybdena	molybdena	
Nickel	nickel	
Gold	gold	
Platina	platina	
Lead	lead	Galena.
Tungstein	tungstein	
Zinc	zinc	Blende.
Potash	potash	Alkaline liver of sulphur with fixed vegetable alkali.
Soda	soda	Alkaline liver of sulphur with fixed mineral alkali.
Ammoniac	ammoniac	Volatile liver of sulphur, smoking liquor of Boyle.
Lime	lime	Calcareous liver of sulphur.
Magnesia	magnesia	Magnesian liver of sulphur.
Barytes	barytes	Barytic liver of sulphur.
Argill	argill	Yet unknown.

[Pg 203]

SECT. VIII.—*Observations on Sulphur, and its Combinations.*

Sulphur is a combustible substance, having a very great tendency to combination; it is naturally in a solid state in the ordinary temperature, and requires a heat somewhat higher than boiling water to make it liquify. Sulphur is formed by nature in a considerable degree of purity in the neighbourhood of volcanos; we find it likewise, chiefly in the state of sulphuric acid, combined with argill in aluminous schistus, with lime in gypsum, &c. From these combinations it may be procured in the state of sulphur, by carrying off its oxygen by means of charcoal in a red heat; carbonic acid is formed, and escapes in the state of gas; the sulphur remains combined with the clay, lime, &c. in the state of sulphuret, which is decomposed by acids; the acid unites with the earth into a neutral salt, and the sulphur is precipitated.[Pg 204]

TABLE *of the Binary Combinations of Phosphorus with the Simple Substances.*

Simple Substances.	Resulting Compounds.
Caloric	Phosphoric gas.
Oxygen	{ Oxyd of phosphorus. { Phosphorous acid. { Phosphoric acid.
Hydrogen	Phosphuret of hydrogen.
Azote	Phosphuret of azote.
Sulphur	Phosphuret of Sulphur.
Charcoal	Phosphuret of charcoal.
Metallic substances	Phosphuret of metals(A).
Potash} Soda} Ammoniac } Lime} Barytes} Magnesia} Argill}	Phosphuret of Potash, Soda, &c.(B)

[Note A: Of all these combinations of phosphorus with metals, that with iron only is hitherto known, forming the substance formerly called Siderite; neither is it yet ascertained whether, in this combination, the phosphorus be oxygenated or not.—A.]

[Note B: These combinations of phosphorus with the alkalies and earths are not yet known; and, from the experiments of Mr Gengembre, they appear to be impossible—A.][Pg 205]

SECT. IX.—*Observations upon Phosphorus, and its Combinations.*

Phosphorus is a simple combustible substance, which was unknown to chemists till 1667, when it was discovered by Brandt, who kept the process secret; soon after Kunkel found out Brandt's method of preparation, and made it public. It has been ever since known by the name of Kunkel's phosphorus. It was for a long time procured only from urine; and, though Homberg gave an account of the process in the Memoirs of the Academy for 1692, all the philosophers of Europe were supplied with it from England. It was first made in France in 1737, before a committee of the Academy at the Royal Garden. At present it is procured in a more commodious and more oeconomical manner from animal bones, which are real calcareous phosphats, according to the process of Messrs Gahn, Scheele, Rouelle, &c. The bones of adult animals being calcined to whiteness, are pounded, and passed through a fine silk sieve; pour upon the fine powder a quantity of dilute sulphuric acid, less than is sufficient for dissolving the whole. This acid unites with the calcareous earth of the bones into a sulphat of lime, and the phosphoric acid remains free in the liquor. The liquid[Pg 206] is decanted off, and the residuum washed with boiling water; this water which has been used to wash out the adhering acid is joined with what was before decanted off, and the whole is gradually evaporated; the dissolved sulphat of lime cristallizes in form of silky threads, which are removed, and by continuing the evaporation we procure the phosphoric acid under the appearance of a white pellucid glass. When this is powdered, and mixed with one third its weight of charcoal, we procure very pure phosphorus by sublimation. The phosphoric acid, as procured by the above process, is never so pure as that obtained by oxygenating pure phosphorus either by combustion or by means of nitric acid; wherefore this latter should always be employed in experiments of research.

Phosphorus is found in almost all animal substances, and in some plants which give a kind of animal analysis. In all these it is usually combined with charcoal, hydrogen, and azote, forming very compound radicals, which are, for the most part, in the state of oxyds by a first degree of union with oxygen. The discovery of Mr Hassenfratz, of phosphorus being contained in charcoal, gives reason to suspect that it is more common in the vegetable kingdom than has generally been supposed: It is certain, that, by proper processes, it may be procured from every individual of some of the families of plants.[Pg 207]

As no experiment has hitherto given reason to suspect that phosphorus is a compound body, I have arranged it with the simple or elementary substances. It takes fire at the temperature of 32° (104°) of the thermometer.

TABLE *of the Binary Combinations of Charcoal.*

Simple Substances.	*Resulting Compounds.*	
Oxygen	{Oxyd of charcoal	Unknown.
	{Carbonic acid	Fixed air, chalky acid.
Sulphur	Carburet of sulphur}	
Phosphorus	Carburet of phosphorus}	Unknown.
Azote	Carburet of azote}	
Hydrogen	{Carbono-hydrous radical	
	{Fixed and volatile oils	
		{Of these only the

Metallic substances	Carburets of metals	carburets of {iron and zinc are known, and {were formerly called Plumbago.
Alkalies and earths	Carburet of potash, &c.	Unknown.

[Pg 208]
SECT. X.—*Observations upon Charcoal, and its Combinations with Simple Substances.*

As charcoal has not been hitherto decomposed, it must, in the present state of our knowledge, be considered as a simple substance. By modern experiments it appears to exist ready formed in vegetables; and I have already remarked, that, in these, it is combined with hydrogen, sometimes with azote and phosphorus, forming compound radicals, which may be changed into oxyds or acids according to their degree of oxygenation.

To obtain the charcoal contained in vegetable or animal substances, we subject them to the action of fire, at first moderate, and afterwards very strong, on purpose to drive off the last portions of water, which adhere very obstinately to the charcoal. For chemical purposes, this is usually done in retorts of stone-ware or porcellain, into which the wood, or other matter, is introduced, and then placed in a reverberatory furnace, raised gradually to its greatest heat; the heat volatilizes, or changes into gas, all the parts of the body susceptible of combining with caloric into that form, and the charcoal, being more fixed in its nature, remains in the retort[Pg 209] combined with a little earth and some fixed salts.

In the business of charring wood, this is done by a less expensive process. The wood is disposed in heaps, and covered with earth, so as to prevent the access of any more air than is absolutely necessary for supporting the fire, which is kept up till all the water and oil is driven off, after which the fire is extinguished by shutting up all the air-holes.

We may analyse charcoal either by combustion in air, or rather in oxygen gas, or by means of nitric acid. In either case we convert it into carbonic acid, and sometimes a little potash and some neutral salts remain. This analysis has hitherto been but little attended to by chemists; and we are not even certain if potash exists in charcoal before combustion, or whether it be formed by means of some unknown combination during that process.

SECT. XI.—*Observations upon the Muriatic, Fluoric, and Boracic Radicals, and their Combinations.*

As the combinations of these substances, either with each other, or with the other combustible bodies, are hitherto entirely unknown, we have[Pg 210] not attempted to form any table for their nomenclature. We only know that these radicals are susceptible of oxygenation, and of forming the muriatic, fluoric, and boracic acids, and that in the acid state they enter into a number of combinations, to be afterwards detailed. Chemistry has hitherto been unable to disoxygenate any of them, so as to produce them in a simple state. For this purpose, some substance must be employed to which oxygen has a stronger affinity than to their radicals, either by means of single affinity, or by double elective attraction. All that is known relative to the origin of the radicals of these acids will be mentioned in the sections set apart for considering their combinations with the salifiable bases.

SECT. XII.—*Observations upon the Combinations of Metals with each other.*

Before closing our account of the simple or elementary substances, it might be supposed necessary to give a table of alloys or combinations of metals with each other; but, as such a table would be both exceedingly voluminous and very unsatisfactory, without going into a series of experiments not yet attempted, I have thought it adviseable to omit it altogether. All[Pg 211] that is necessary to be mentioned is, that these alloys should be named according to the metal in largest proportion in the mixture or combination; thus the term *alloy of gold and silver*, or gold alloyed with silver, indicates that gold is the predominating metal.

Metallic alloys, like all other combinations, have a point of saturation. It would even appear, from the experiments of Mr de la Briche, that they have two perfectly distinct degrees of saturation.[Pg 212]

TABLE *of the Combinations of Azote in the state of Nitrous Acid with the Salifiable Bases, arranged according to the affinities of these Bases with the Acid.*

Names of the bases.	Names of the neutral salts.	Notes.
	New nomenclature.	
Barytes	Nitrite of barytes.	{
Potash	potash.	{These salts are only
Soda	soda.	{known of late, and
Lime	lime.	{have received no particular
Magnesia	magnesia.	{name in the old
Ammoniac	ammoniac.	{nomenclature.
Argill	argill.	{
		{As metals dissolve both in nitrous and
Oxyd of zinc	zinc.	{nitric acids, metallic salts must of
iron	iron.	{consequence be formed having
manganese	manganese.	{different degrees of oxygenation.
cobalt	cobalt.	{Those wherein the metal is
nickel	nickel.	{least oxygenated must be
lead	lead.	{called Nitrites, when more so,
tin	tin.	{Nitrats; but the limits of this
copper	copper.	{distinction are difficultly

87

bismuth	bismuth.	{ascertainable. The older
antimony	antimony.	{chemists were not acquainted
arsenic	arsenic.	{with any of these salts.
mercury	mercury.	{

silver	{It is extremely probable that gold, silver
gold	{and platina only form nitrats, and cannot subsist
platina	{in the state of nitrites.

[Pg 213]

TABLE *of the Combinations of Azote, completely saturated with Oxygen, in the state of Nitric Acid, with the Salifiable Bases, in the order of the affinity with the Acid.*

Bases.	Names of the resulting neutral salts.	
	New nomenclature.	Old nomenclature.
Barytes	Nitrat of barytes	Nitre, with a base of heavy earth.
Potash	Nitrat of potash	Nitre, saltpetre. Nitre with base of potash.
Soda	Nitrat of soda	Quadrangular nitre. Nitre with base of mineral alkali.
Lime	Nitrat of lime	Calcareous nitre. Nitre with calcareous base. Mother water of nitre, or saltpetre.
Magnesia	Nitrat of magnesia	Magnesian nitre. Nitre with base of magnesia.
Ammoniac	Nitrat of ammoniac	Ammoniacal nitre.
Argill	Nitrat of argill	Nitrous alum. Argillaceous nitre. Nitre with base of earth of alum.
Oxyd of zinc	Nitrat of zinc	Nitre of zinc.
Oxyd of iron	Nitrat of iron	Nitre of iron. Martial nitre. Nitrated iron.
Oxyd of manganese	Nitrat of manganese	Nitre of manganese.

nganese		nganese	
cobalt		cobalt	Nitre of cobalt.
nickel		nickel	Nitre of nickel.
lead		lead	Saturnine nitre. Nitre of lead.
	tin		tin	Nitre of tin.
copper		copper	Nitre of copper or of Venus.
bismuth		bismuth	Nitre of bismuth.
antimony		antimony	Nitre of antimony.
arsenic		arsenic	Arsenical nitre.
mercury		mercury	Mercurial nitre.
silver		silver	Nitre of silver or luna. Lunar caustic.
gold		gold	Nitre of gold.
platina		platina	Nitre of platina.

[Pg 214]

SECT. XIII.—*Observations upon the Nitrous and Nitric Acids, and their Combinations.*

The nitrous and nitric acids are procured from a neutral salt long known in the arts under the name of *saltpetre*. This salt is extracted by lixiviation from the rubbish of old buildings, from the earth of cellars, stables, or barns, and in general of all inhabited places. In these earths the nitric acid is usually combined with lime and magnesia, sometimes with potash, and rarely with argill. As all these salts, excepting the nitrat of potash, attract the moisture of the air, and consequently would be difficultly preserved, advantage is taken, in the manufactures of saltpetre and the royal refining house, of the greater affinity of the nitric acid to potash than these other bases, by which means the lime, magnesia, and argill, are precipitated, and all these nitrats are reduced to the nitrat of potash or saltpetre[41].

The nitric acid is procured from this salt by distillation, from three parts of pure saltpetre decomposed by one part of concentrated sulphuric[Pg 215] acid, in a retort with Woulfe's apparatus, (Pl. IV. fig. 1.) having its bottles half filled with water, and all its joints carefully luted. The nitrous acid passes over in form of red vapours surcharged with nitrous gas, or, in other words, not saturated with oxygen. Part of the acid condenses in the recipient in form of a dark orange red liquid, while the rest combines with the water in the bottles. During the distillation, a large quantity of oxygen gas escapes, owing to the greater affinity of oxygen to caloric, in a high temperature, than to nitrous acid, though in the usual temperature of the atmosphere this affinity is reversed. It is from the disengagement of oxygen that the nitric acid of the neutral salt is in this operation converted into nitrous acid. It is brought back to the state of nitric acid by heating over a gentle fire, which drives off the superabundant nitrous gas, and leaves the nitric acid much diluted with water.

Nitric acid is procurable in a more concentrated state, and with much less loss, by mixing very dry clay with saltpetre. This mixture is put into an earthern retort, and distilled with a strong fire. The clay combines with the potash, for which it has great affinity, and the nitric acid passes over, slightly impregnated with nitrous gas. This is easily disengaged by heating the acid gently in a retort, a small quantity[Pg 216] of nitrous gas passes over into the recipient, and very pure concentrated nitric acid remains in the retort.

We have already seen that azote is the nitric radical. If to 20-1/2 parts, by weight, of azote 43-1/2 parts of oxygen be added, 64 parts of nitrous gas are formed; and, if to this we join 36 additional parts of oxygen, 100 parts of nitric acid result from the combination. Intermediate quantities of oxygen between these two extremes of oxygenation produce different species of nitrous acid, or, in other words, nitric acid less or more impregnated with nitrous gas. I ascertained the above proportions by means of decomposition; and, though I cannot answer for their absolute accuracy, they cannot be far removed from truth. Mr Cavendish, who first showed by synthetic experiments that azote is the base of nitric acid, gives the proportions of azote a little larger than I have done; but, as it is not improbable that he produced the nitrous acid and not the nitric, that circumstance explains in some degree the difference in the results of our experiments.

As, in all experiments of a philosophical nature, the utmost possible degree of accuracy is required, we must procure the nitric acid for experimental purposes, from nitre which has been previously purified from all foreign matter. If, after distillation, any sulphuric acid is suspected[Pg 217] in the nitric acid, it is easily separated by dropping in a little nitrat of barytes, so long as any precipitation takes place; the sulphuric acid, from its greater affinity, attracts the barytes, and forms with it an insoluble neutral salt, which falls to the bottom. It may be purified in the same manner from muriatic acid, by dropping in a little nitrat of silver so long as any precipitation of muriat of silver is produced. When these two precipitations are finished, distill off about seven-eighths of the acid by a gentle heat, and what comes over is in the most perfect degree of purity.

The nitric acid is one of the most prone to combination, and is at the same time very easily decomposed. Almost all the simple substances, with the exception of gold, silver, and platina, rob it less or more of its oxygen; some of them even decompose it altogether. It was very anciently known, and its combinations have been more studied by chemists than those of any other acid. These combinations were named *nitres* by Messrs Macquer and Beaumé; but we have changed their names to nitrats and nitrites, according as they are formed by nitric or by nitrous acid, and have added the specific name of each particular base, to distinguish the several combinations from each other.[Pg 218]

TABLE *of the Combinations of Sulphuric Acid with the Salifiable Bases, in the order of affinity.*

Names of the bases.	Resulting compounds.	
	New nomenclature.	Old nomenclature.
Barytes	Sulphat of barytes	Heavy spar. Vitriol of heavy earth.
Potash	Sulphat of potash	Vitriolated tartar. Sal de duobus. Arcanum duplicatam.
Soda	Sulphat of soda	Glauber's salt.

Lime		lime	Selenite, gypsum, calcareous vitriol.
Magnesia		magnesia	Epsom salt, sedlitz salt, magnesian vitriol.
Ammoniac		ammoniac	Glauber's secret sal ammoniac.
Argill		argill	Alum.
Oxyd of	zinc	zinc	White vitriol, goslar vitriol, white coperas, vitriol of zinc.
	iron	iron	Green coperas, green vitriol, martial vitriol, vitriol of iron.
	manganese	manganese	Vitriol of manganese.
	cobalt	cobalt	Vitriol of cobalt.
	nickel	nickel	Vitriol of nickel.
	lead	lead	Vitriol of lead.
	tin	tin	Vitriol of tin.
	copper	copper	Blue coperas, blue vitriol, Roman vitriol, vitriol of copper.
	bismuth	bismuth	Vitriol of bismuth.
	antimony	antimony	Vitriol of antimony.
	arsenic	arsenic	Vitriol of arsenic.
	mercury	mercury	Vitriol of mercury.
	silver	silver	Vitriol of silver.
	gold	gold	Vitriol of gold.
	platina	platina	Vitriol of platina.

[Pg 219]

SECT. XIV.—*Observations upon Sulphuric Acid and its Combinations.*

For a long time this acid was procured by distillation from sulphat of iron, in which sulphuric acid and oxyd of iron are combined, according to the process described by Basil Valentine in the fifteenth century; but, in modern times, it is procured more oeconomically by the combustion of sulphur in proper vessels. Both to facilitate the combustion, and to assist the oxygenation of the sulphur, a little powdered saltpetre, nitrat of potash, is mixed with it; the nitre is decomposed, and gives out its oxygen to the sulphur, which contributes to its conversion into acid. Notwithstanding this addition, the sulphur will only continue to burn in close vessels for a limited time; the combination ceases, because the oxygen is exhausted, and the air of the vessels reduced almost to pure azotic gas, and because the acid itself remains long in the state of vapour, and hinders the progress of combustion.

In the manufactories for making sulphuric acid in the large way, the mixture of nitre and sulphur is burnt in large close built chambers lined with lead, having a little water at the bottom for facilitating the condensation of the vapours.[Pg 220] Afterwards, by distillation in large retorts with a gentle heat, the water passes over, slightly impregnated with acid, and the sulphuric acid remains behind in a concentrated state. It is then pellucid, without any flavour, and nearly double the weight of an equal bulk of water. This process would be greatly facilitated, and the combustion much prolonged, by introducing fresh air into the chambers, by means of several pairs of bellows directed towards the flame of the sulphur, and by allowing the nitrous gas to escape through long serpentine canals, in contact with water, to absorb any sulphuric or sulphurous acid gas it might contain.

By one experiment, Mr Berthollet found that 69 parts of sulphur in combustion, united with 31 parts of oxygen, to form 100 parts of sulphuric acid; and, by another experiment, made in a different manner, he calculates that 100 parts of sulphuric acid consists of 72 parts sulphur, combined with 28 parts of oxygen, all by weight.

This acid, in common with every other, can only dissolve metals when they have been previously oxydated; but most of the metals are capable of decomposing a part of the acid, so as to carry off a sufficient quantity of oxygen, to render themselves soluble in the part of the acid which remains undecomposed. This happens with silver, mercury, iron, and zinc, in boiling[Pg 221] concentrated sulphuric acid; they become first oxydated by decomposing part of the acid, and then dissolve in the other part; but they do not sufficiently disoxygenate the decomposed part of the acid to reconvert it into sulphur; it is only reduced to the state of sulphurous acid, which, being volatilised by the heat, flies off in form of sulphurous acid gas.

Silver, mercury, and all the other metals except iron and zinc, are insoluble in diluted sulphuric acid, because they have not sufficient affinity with oxygen to draw it off from its combination either with the sulphur, the sulphurous acid, or the hydrogen; but iron and zinc, being assisted by the action of the acid, decompose the water, and become oxydated at its expence, without the help of heat.[Pg 222]

TABLE *of the Combinations of the Sulphurous Acid with the Salifiable Bases, in the order of affinity.*

Names of the Bases.	Names of the Neutral Salts.
Barytes	Sulphite of barytes.
Potash	Potash.
Soda	soda.
Lime	lime
Ma	mag

gnesia	nesia.
Ammoniac	ammoniac.
Argill	argil.
Oxyd of zinc	zinc.
iron	iron.
manganese	manganese.
cobalt	cobalt.
nickel	nickel.
lead	lead.
tin	tin.
copper	copper.
bismuth	bismuth.
antimony	antimony.
arsenic	arsenic.
mercury	mercury.
silver	silver.
gold	gold.
platina	platina.

Note.—The only one of these salts known to the old chemists was the sulphite of potash, under the name of *Stahl's sulphureous salt*. So that, before our new nomenclature, these compounds must have been named *Stahl's sulphureous salt*, having base of fixed vegetable alkali, and so of the rest.

In this Table we have followed Bergman's order of affinity of the sulphuric acid, which is the same in regard to the earths and alkalies, but it is not certain if the order be the same for the metallic oxyds.—A.[Pg 223]

SECT. XV.—*Observations upon Sulphurous Acid, and its Combinations.*

The sulphurous acid is formed by the union of oxygen with sulphur by a lesser degree of oxygenation than the sulphuric acid. It is procurable either by burning sulphur slowly, or by distilling sulphuric acid from silver, antimony, lead, mercury, or charcoal; by which operation a part of the oxygen quits the acid, and unites to these oxydable bases, and the acid passes over in the sulphurous state of oxygenation. This acid, in the common pressure and temperature of the air, can only exist in form of gas; but it appears, from the experiments of Mr Clouet, that, in a very low temperature, it condenses, and becomes fluid. Water absorbs a great deal more of this gas than of carbonic acid gas, but much less than it does of muriatic acid gas.

That the metals cannot be dissolved in acids without being previously oxydated, or by procuring oxygen, for that purpose, from the acids during solution, is a general and well established fact, which I have perhaps repeated too often. Hence, as sulphurous acid is already deprived of great part of the oxygen necessary for forming the sulphuric acid, it is more disposed[Pg 224] to recover oxygen, than to furnish it to the greatest part of the metals; and, for this reason, it cannot dissolve them, unless previously oxydated by other means. From the same principle it is that the metallic oxyds dissolve without effervescence, and with great facility, in sulphurous acid. This acid, like the muriatic, has even the property of dissolving metallic oxyds surcharged with oxygen, and consequently insoluble in sulphuric acid, and in this way forms true sulphats. Hence we might be led to conclude that there are no metallic sulphites, were it not that the phenomena which accompany the solution of iron, mercury, and some other metals, convince us that these metallic substances are susceptible of two degrees of oxydation, during their solution in acids. Hence the neutral salt in which the metal is least oxydated must be named *sulphite*, and that in which it is fully oxydated must be called *sulphat*. It is yet unknown whether this distinction is applicable to any of the metallic sulphats, except those of iron and mercury.[Pg 225]

TABLE *of the Combinations of Phosphorous and Phosphoric Acids, with the Salifiable Bases, in the Order of Affinity.*

Names of the Bases.	Names of the Neutral Salts formed by	
	Phosphorous Acid,	Phosphoric Acid.
	Phosphites of(B)	Phosphats of(C)
Lime	lime	lime.
Barytes	barytes	barytes.
Magnesia	magnesia	magnesia
Potash	potash	potash.
Soda	soda	soda.
Ammoniac	ammoniac	ammoniac.
Argill	argill	argill.

94

of(A)	Oxyds		
	zinc	zinc	zinc.
	iron	iron	iron.
	manganese	manganese	manganese.
	cobalt	cobalt	cobalt.
	nickel	nickel	nickel.
	lead	lead	lead.
	tin	tin	tin.
	copper	copper	copper.
	bismuth	bismuth	bismuth.
	antimony	antimony	antimony.
	arsenic	arsenic	arsenic.
	mercury	mercury	mercury.
	silver	silver	silver.
	gold	gold	gold.
	platina	platina	platina.

[Note A: The existence of metallic phosphites supposes that metals are susceptible of solution in phosphoric acid at different degrees of oxygenation, which is not yet ascertained.—A.]

[Note B: All the phosphites were unknown till lately, and consequently have not hitherto received names.—A.]

[Note C: The greater part of the phosphats were only discovered of late, and have not yet been named.—A.][Pg 226]

SECT. XVI.—*Observations upon Phosphorous and Phosphoric Acids, and their Combinations.*

Under the article Phosphorus, Part II. Sect. X. we have already given a history of the discovery of that singular substance, with some observations upon the mode of its existence in vegetable and animal bodies. The best method of obtaining this acid in a state of purity is by burning well purified phosphorus under bell-glasses, moistened on the inside with distilled water; during combustion it absorbs twice and a half its weight of oxygen; so that 100 parts of phosphoric acid is composed of 28-1/2 parts of phosphorus united to 71-1/2 parts of oxygen. This acid may be obtained concrete, in form of white flakes, which greedily attract the moisture of the air, by burning phosphorus in a dry glass over mercury.

To obtain phosphorous acid, which is phosphorus less oxygenated than in the state of phosphoric acid, the phosphorus must be burnt by a very slow spontaneous combustion over a glass-funnel leading into a crystal phial; after a few days, the phosphorus is found oxygenated, and the phosphorous acid, in proportion as it forms, has attracted moisture from the air, and dropped into the phial. The phosphorous[Pg 227] acid is readily changed into phosphoric acid by exposure for a long time to the free air; it absorbs oxygen from the air, and becomes fully oxygenated.

As phosphorus has a sufficient affinity for oxygen to attract it from the nitric and muriatic acids, we may form phosphoric acid, by means of these acids, in a very simple and cheap manner. Fill a tubulated receiver, half full of concentrated nitric acid, and heat it gently, then throw in small pieces of phosphorus through the tube, these are dissolved with effervescence and red fumes of nitrous gas fly off; add phosphorus so long as it will dissolve, and then increase the fire under the retort to drive off the last particles of nitric acid; phosphoric acid, partly fluid and partly concrete, remains in the retort.[Pg 228]

TABLE *of the Combinations of Carbonic Acid, with the Salifiable Bases, in the Order of Affinity.*

Names of Bases	Resulting Neutral Salts.	
	New Nomenclature.	Old Nomenclature.
Barytes	Carbonates of barytes(A)	Aërated or effervescent heavy earth.
Lime	lime	Chalk, calcareous spar, Aërated calcareous earth.
Potash	potash	Effervescing or aërated fixed vegetable alkali, mephitis of potash.
Soda	soda	Aërated or effervescing fixed mineral alkali, mephitic soda.
Magnesia	magnesia	Aërated, effervescing, mild, or mephitic magnesia.
Ammoniac	ammoniac	Aërated, effervescing, mild, or mephitic volatile alkali.
Argill	argill	Aërated or effervescing argillaceous earth, or earth of alum.
Oxyds of		
zinc	zinc	Zinc spar, mephitic or aërated zinc.
iron	iron	Sparry iron-ore, mephitic or aërated iron.
manganese	manganese	Aërated manganese.
cobalt	cobalt	Aërated cobalt.
nickel	nickel	Aërated nickel.
lead	lead	Sparry lead-ore, or aërated lead.

	d		
tin		tin	Aërated tin.
copper		copper	Aërated copper.
bismuth		bismuth	Aërated bismuth.
antimony		antimony	Aërated antimony.
arsenic		arsenic	Aërated arsenic.
mercury		mercury	Aërated mercury.
silver		silver	Aërated silver.
gold	d	gold	Aërated gold.
platina		platina	Aërated platina.

[Note A: As these salts have only been understood of late, they have not, properly speaking, any old names. Mr Morveau, in the First Volume of the Encyclopedia, calls them *Mephites*; Mr Bergman gives them the name of *aërated*; and Mr de Fourcroy, who calls the carbonic acid *chalky acid*, gives them the name of *chalks*.—A][Pg 229]

SECT. XVII.—*Observations upon Carbonic Acid, and its Combinations.*

Of all the known acids, the carbonic is the most abundant in nature; it exists ready formed in chalk, marble, and all the calcareous stones, in which it is neutralized by a particular earth called *lime*. To disengage it from this combination, nothing more is requisite than to add some sulphuric acid, or any other which has a stronger affinity for lime; a brisk effervescence ensues, which is produced by the disengagement of the carbonic acid which assumes the state of gas immediately upon being set free. This gas, incapable of being condensed into the solid or liquid form by any degree of cold or of pressure hitherto known, unites to about its own bulk of water, and thereby forms a very weak acid. It may likewise be obtained in great abundance from saccharine matter in fermentation, but is then contaminated by a small portion of alkohol which it holds in solution.

As charcoal is the radical of this acid, we may form it artificially, by burning charcoal in oxygen gas, or by combining charcoal and metallic oxyds in proper proportions; the oxygen of the oxyd combines with the charcoal, forming[Pg 230] carbonic acid gas, and the metal being left free, recovers its metallic or reguline form.

We are indebted for our first knowledge of this acid to Dr Black, before whose time its property of remaining always in the state of gas had made it to elude the researches of chemistry.

It would be a most valuable discovery to society, if we could decompose this gas by any cheap process, as by that means we might obtain, for economical purposes, the immense store of charcoal contained in calcareous earths, marbles, limestones, &c. This cannot be effected by single affinity, because, to decompose the carbonic acid, it requires a substance as combustible as charcoal itself, so that we should only make an exchange of one combustible body for another not more valuable; but it may possibly be accomplished by double affinity, since this process is so readily performed by Nature, during vegetation, from the most common materials.[Pg 231]

TABLE *of the Combinations of Muriatic Acid, with the Salifiable Bases, in the Order of Affinity.*

Names of the bases.	Resulting Neutral Salts.		
		New nomenclature.	Old nomenclature.
	Barytes.	Muriat of barytes	Sea-salt, having base of heavy earth.
	Potash	potash	Febrifuge salt of Sylvius: Muriated vegetable fixed alkali.
	Soda	soda	Sea-salt.
	Lime	lime	Muriated lime. Oil of lime.
a	Magnesia	magnesia	Marine Epsom salt. Muriated magnesia.
iac	Ammon	ammoniac	Sal ammoniac.
	Argill	argill	{Muriated alum, sea-salt with base of earth of alum.
	Oxyd of		
	zinc	zinc	Sea-salt of, or muriatic zinc.
	iron	iron	Salt of iron, Martial sea-salt.
	manganese	manganese	Sea-salt of manganese.
	cobalt	cobalt	Sea-salt of cobalt.
	nickel	nickel	Sea-salt of nickel.
	lead	lead	Horny-lead. Plumbum corneum.
	tin	smoaking of tin solid of tin	Smoaking liquor of Libavius. Solid butter of tin.
	copper	copper	Sea-salt of copper.
	bismuth	bismuth	Sea-salt of bismuth.
	antimony	antimony	Sea-salt of antimony.
	arsenic	arsenic	Sea-salt of arsenic.
	mercury	{sweet of	Sweet sublimate of mercury, calomel,

	mercury	aquila alba.
	{corrosive of mercury	Corrosive sublimate of mercury.
silver	silver	Horny silver, argentum corneum, luna cornea.
gold	gold	Sea-salt of gold.
platina	platina	Sea-salt of platina.

[Pg 232]
TABLE *Of the Combinations of Oxygenated Muriatic Acid, with the Salifiable Bases, in the Order of Affinity.*

Names of the Bases.	Names of the Neutral Salts by the new Nomenclature.
	Oxygenated muriat of
Barytes	barytes.
Potash	potash.
Soda	soda.
Lime	lime.
Magnesia	magnesia.
Argill	argill.
Oxyd of	
zinc	zinc.
iron	iron.
manganese	manganese.
cobalt	cobalt.
nickel	nickel.
lead	lead.
tin	tin.
copper	copper.
bismuth	bismuth.
antimony	antimony.
arsenic	arsenic.

mercury	mercury.
silver	silver.
gold	gold.
platina	platina.

This order of salts, entirely unknown to the ancient chemists, was discovered in 1786 by Mr Berthollet.—A.[Pg 233]

SECT. XIX.—*Observations upon Muriatic and Oxygenated Muriatic Acids, and their Combinations.*

Muriatic acid is very abundant in the mineral kingdom naturally combined with different salifiable bases, especially with soda, lime, and magnesia. In sea-water, and the water of several lakes, it is combined with these three bases, and in mines of rock-salt it is chiefly united to soda. This acid does not appear to have been hitherto decomposed in any chemical experiment; so that we have no idea whatever of the nature of its radical, and only conclude, from analogy with the other acids, that it contains oxygen as its acidifying principle. Mr Berthollet suspects the radical to be of a metallic nature; but, as Nature appears to form this acid daily, in inhabited places, by combining miasmata with aëriform fluids, this must necessarily suppose a metallic gas to exist in the atmosphere, which is certainly not impossible, but cannot be admitted without proof.

The muriatic acid has only a moderate adherence to the salifiable bases, and can readily be driven from its combination with these by sulphuric acid. Other acids, as the nitric, for instance, may answer the same purpose; but nitric acid being volatile, would mix, during distillation,[Pg 234] with the muriatic. About one part of sulphuric acid is sufficient to decompose two parts of decrepitated sea-salt. This operation is performed in a tubulated retort, having Woulfe's apparatus, (Pl. IV. Fig. 1.), adapted to it. When all the junctures are properly lured, the sea-salt is put into the retort through the tube, the sulphuric acid is poured on, and the opening immediately closed with its ground crystal stopper. As the muriatic acid can only subsist in the gaseous form in the ordinary temperature, we could not condense it without the presence of water. Hence the use of the water with which the bottles in Woulfe's apparatus are half filled; the muriatic acid gas, driven off from the sea-salt in the retort, combines with the water, and forms what the old chemists called *smoaking spirit of salt*, or *Glauber's spirit of sea-salt*, which we now name *muriatic acid*.

The acid obtained by the above process is still capable of combining with a farther dose of oxygen, by being distilled from the oxyds of manganese, lead, or mercury, and the resulting acid, which we name *oxygenated muriatic acid*, can only, like the former, exist in the gasseous form, and is absorbed, in a much smaller quantity by water. When the impregnation of water with this gas is pushed beyond a certain point, the superabundant acid precipitates to the bottom of the vessels in a concrete form. Mr Berthollet has[Pg 235] shown that this acid is capable of combining with a great number of the salifiable bases; the neutral salts which result from this union are susceptible of deflagrating with charcoal, and many of the metallic substances; these deflagrations are very violent and dangerous, owing to the great quantity of caloric which the oxygen carries alongst with it into the composition of oxygenated muriatic acid.[Pg 236]

TABLE *of the Combinations of Nitro-muriatic Acid with the Salifiable Bases, in the Order of Affinity, so far as is known.*

	Names of the Bases.		Names of the Neutral Salts.		
ill	Arg		Nitro-muriat of	l.	argil
	Am				am

moniac		moniac.
Oxyd of antimony		antimony.
silver		silver.
arsenic		arsenic.
Barytes		barytes.
Oxyd of bismuth		bismuth.
Lime		lime.
Oxyd of cobalt		cobalt.
copper		copper.
tin		tin.
iron		iron.
Magnesia		magnesia.
Oxyd of manganese		manganese.
mercury		mercury.
molybdena		molybdena.
nickel		nickel.
gold		gold.

Names of the Bases.	Names of the Neutral Salts.
platina	platina.
lead	lead.
Potash	potash.
Soda	soda.
Oxyd of tungstein	tungstein.
zinc	zinc.

Note.—Most of these combinations, especially those with the earths and alkalies, have been little examined, and we are yet to learn whether they form a mixed salt in which the compound radical remains combined, or if the two acids separate, to form two distinct neutral salts.—A.[Pg 237]

SECT. XX.—*Observations upon the Nitro-Muriatic Acid, and its Combinations.*

The nitro-muriatic acid, formerly called *aqua regia*, is formed by a mixture of nitric and muriatic acids; the radicals of these two acids combine together, and form a compound base, from which an acid is produced, having properties peculiar to itself, and distinct from those of all other acids, especially the property of dissolving gold and platina.

In dissolutions of metals in this acid, as in all other acids, the metals are first oxydated by attracting a part of the oxygen from the compound radical. This occasions a disengagement of a particular species of gas not hitherto described, which may be called *nitro-muriatic gas*; it has a very disagreeable smell, and is fatal to animal life when respired; it attacks iron, and causes it to rust; it is absorbed in considerable quantity by water, which thereby acquires some slight characters of acidity. I had occasion to make these remarks during a course of experiments upon platina, in which I dissolved a considerable quantity of that metal in nitro-muriatic acid.

I at first suspected that, in the mixture of nitric and muriatic acids, the latter attracted a[Pg 238]part of the oxygen from the former, and became converted into oxygenated muriatic acid, which gave it the property of dissolving gold; but several facts remain inexplicable upon this supposition. Were it so, we must be able to disengage nitrous gas by heating this acid, which however does not sensibly happen. From these considerations, I am led to adopt the opinion of Mr Berthollet, and to consider nitro-muriatic acid as a single acid, with a compound base or radical.[Pg 239]

TABLE of the Combinations of Fluoric Acid, with the Salifiable Bases, in the Order of Affinity.

Names of the Bases.	Names of the Neutral Salts.
Lime	Fluat of lime.
Barytes	barytes.
Magn	magn

gnesia	esia.
Potash	potash.
Soda	soda.
Ammoniac	ammoniac.
Oxyd of	
zinc	zinc.
manganese	manganese.
iron	iron.
lead	lead.
tin	tin.
cobalt	cobalt.
copper	copper.
nickel	nickel.
arsenic	arsenic.
bismuth	bismuth.
mercury	mercury.
silver	silver.
gold	gold.
platina	platina.

And by the dry way,

Argill	Fluat of	argill.

Note.—These combinations were entirely unknown to the old chemists, and consequently have no names in the old nomenclature.—A.[Pg 240]

SECT. XXI.—*Observations upon the Fluoric Acid, and its Combinations.*

Fluoric exists ready formed by Nature in the fluoric spars[42], combined with calcareous earth, so as to form an insoluble neutral salt. To obtain it disengaged from that combination, fluor spar, or fluat of lime, is put into a leaden retort, with a proper quantity of sulphuric acid, a recipient likewise of lead, half full of water, is adapted, and fire is applied to the retort. The sulphuric acid, from its greater affinity, expels the fluoric acid which passes over and is absorbed by the water in the receiver. As fluoric acid is naturally in the gasseous form in the ordinary temperature, we can receive it in a pneumato-chemical apparatus over mercury. We are obliged to employ metallic vessels in this process, because fluoric acid dissolves glass and silicious earth, and even renders these bodies volatile, carrying them over with itself in distillation in the gasseous form.

We are indebted to Mr Margraff for our first acquaintance with this acid, though, as he could never procure it free from combination with a considerable quantity of silicious earth, he was[Pg 241] ignorant of its being an acid sui generis. The Duke de Liancourt, under the name of Mr Boulanger, considerably increased our knowledge of its properties; and Mr Scheele seems to have exhausted the subject. The only thing remaining is to endeavour to discover the nature of the fluoric radical, of which we cannot hitherto form any ideas, as the acid does not appear to have been decomposed in any experiment. It is only by means of compound affinity that experiments ought to be made with this view, with any probability of success.[Pg 242]

TABLE *of the Combinations of Boracic Acid, with the Salifiable Bases, in the Order of Affinity.*

Bases.	Neutral Salts.
Lime	Borat of lime.
Barytes	barytes.
Magnesia	magnesia.
Potash	potash.
Soda	soda.
Ammoniac	ammoniac.
Oxyd of zinc	zinc.
iron	iron.

lead	lead.
tin	tin.
cobalt	cobalt.
copper	copper.
nickel	nickel.
mercury	mercury.
Argill	argill.

Note.—Most of these combinations were neither known nor named by the old chemists. The boracic acid was formerly called *sedative salt*, and its compounds *borax*, with base of fixed vegetable alkali, &c.—A.[Pg 243]

SECT. XXII.—*Observations upon Boracic Acid and its Combinations.*

This is a concrete acid, extracted from a salt procured from India called *borax* or *tincall*. Although borax has been very long employed in the arts, we have as yet very imperfect knowledge of its origin, and of the methods by which it is extracted and purified; there is reason to believe it to be a native salt, found in the earth in certain parts of the east, and in the water of some lakes. The whole trade of borax is in the hands of the Dutch, who have been exclusively possessed of the art of purifying it till very lately, that Messrs L'Eguillier of Paris have rivalled them in the manufacture; but the process still remains a secret to the world.

By chemical analysis we learn that borax is a neutral salt with excess of base, consisting of soda, partly saturated with a peculiar acid long called *Homberg's sedative salt*, now *the boracic acid*. This acid is found in an uncombined state in the waters of certain lakes. That of Cherchiais in Italy contains 94-1/2 grains in each pint of water.

To obtain boracic acid, dissolve some borax in boiling water, filtrate the solution, and add sulphuric acid, or any other having greater affinity[Pg 244] to soda than the boracic acid; this latter acid is separated, and is procured in a crystalline form by cooling. This acid was long considered as being formed during the process by which it is obtained, and was consequently supposed to differ according to the nature of the acid employed in separating it from the soda; but it is now universally acknowledged that it is identically the same acid, in whatever way procured, provided it be properly purified from mixture of other acids, by warning, and by repeated solution and cristallization. It is soluble both in water and alkohol, and has the property of communicating a green colour to the flame of that spirit. This circumstance led to a suspicion of its containing copper, which is not confirmed by any decisive experiment. On the contrary, if it contain any of that metal, it must only be considered as an accidental mixture. It combines with the salifiable bases in the humid way; and though, in this manner, it is incapable of dissolving any of the metals directly, this combination is readily affected by compound affinity.

The Table presents its combinations in the order of affinity in the humid way; but there is a considerable change in the order when we operate via sicca; for, in that case, argill, though the last in our list, must be placed immediately after soda.[Pg 245]

The boracic radical is hitherto unknown; no experiments having as yet been able to decompose the acid; We conclude, from analogy with the other acids, that oxygen exists in its composition as the acidifying principle.[Pg 246]

TABLE *of the Combinations of Arseniac Acid, with the Salifiable Bases, in the Order of Affinity.*

Bases.	Neutral Salts.
Lime	Arseniat of lime.
Barytes	barytes.
Magnesia	magnesia.
Potash	potash.
Soda	soda.
Ammoniac	ammoniac.
Oxyd of	
zinc	zinc.
manganese	manganese.
iron	iron.
lead	lead.
tin	tin.
cobalt	cobalt.
copper	copper.
nickel	nickel.
bismuth	bismuth.
mercury	mercury.

	anti mony		anti mony.
	silv er		silve r.
	gol d		gold .
	plat ina		plati na.
Arg ill		argil l.	

Note.—This order of salts was entirely unknown to the antient chemists. Mr Macquer, in 1746, discovered the combinations of arseniac acid with potash and soda, to which he gave the name of *arsenical neutral salts.*—A.[Pg 247]

SECT. XXIII.—*Observations upon Arseniac Acid, and its Combinations.*

In the Collections of the Academy for 1746, Mr Macquer shows that, when a mixture of white oxyd of arsenic and nitre are subjected to the action of a strong fire, a neutral salt is obtained, which he calls *neutral salt of arsenic.* At that time, the cause of this singular phenomenon, in which a metal acts the part of an acid, was quite unknown; but more modern experiments teach that, during this process, the arsenic becomes oxygenated, by carrying off the oxygen of the nitric acid; it is thus converted into a real acid, and combines with the potash. There are other methods now known for oxygenating arsenic, and obtaining its acid free from combination. The most simple and most effectual of these is as follows: Dissolve white oxyd of arsenic in three parts, by weight, of muriatic acid; to this solution, in a boiling state, add two parts of nitric acid, and evaporate to dryness. In this process the nitric acid is decomposed, its oxygen unites with the oxyd of arsenic, and converts it into an acid, and the nitrous radical flies off in the state of nitrous gas; whilst the muriatic acid is converted by the heat into muriatic acid gas, and may be collected in proper vessels. The arseniac acid is entirely[Pg 248] freed from the other acids employed during the process by heating it in a crucible till it begins to grow red; what remains is pure concrete arseniac acid.

Mr Scheele's process, which was repeated with great success by Mr Morveau, in the laboratory at Dijon, is as follows: Distil muriatic acid from the black oxyd of manganese, this converts it into oxygenated muriatic acid, by carrying off the oxygen from the manganese, receive this in a recipient containing white oxyd of arsenic, covered by a little distilled water; the arsenic decomposes the oxygenated muriatic acid, by carrying off its supersaturation of oxygen, the arsenic is converted into arseniac acid, and the oxygenated muriatic acid is brought back to the state of common muriatic acid. The two acids are separated by distillation, with a gentle heat increased towards the end of the operation, the muriatic acid passes over, and the arseniac acid remains behind in a white concrete form.

The arseniac acid is considerably less volatile than white oxyd of arsenic; it often contains white oxyd of arsenic in solution, owing to its not being sufficiently oxygenated; this is prevented by continuing to add nitrous acid, as in the former process, till no more nitrous gas is produced. From all these observations I would give the following definition of[Pg 249] arseniac acid. It is a white concrete metallic acid, formed by the combination of arsenic with oxygen, fixed in a red heat, soluble in water, and capable of combining with many of the salifiable bases.

SECT. XXIV.—*Observations upon Molybdic Acid, and its Combinations with Acidifiable Bases*[43].

Molybdena is a particular metallic body, capable of being oxygenated, so far as to become a true concrete acid[44]. For this purpose, one part ore of molybdena, which is a natural sulphuret of that metal, is put into a retort, with five or six parts nitric acid, diluted with a quarter of its weight of water, and heat is applied to the retort; the oxygen of the nitric acid acts both upon the molybdena and the sulphur, converting the one into molybdic, and the other into sulphuric acid; pour on fresh quantities of nitric acid so long as any red fumes of nitrous[Pg 250] gas escape; the molydbena is then oxygenated as far as is possible, and is found at the bottom of the retort in a pulverulent form, resembling chalk. It must be washed in warm water, to separate any adhering particles of sulphuric acid; and, as it is hardly soluble, we lose very little of it in this operation. All its combinations with salifiable bases were unknown to the ancient chemists.

[Pg 251]

TABLE *of the Combinations of Tungstic Acid with the Salifiable Bases.*

Bases.	Neutral Salts.
Lime	Tungstat of lime.
Barytes	barytes.
Magnesia	magnesia.
Potash	potash.
Soda	soda.
Ammoniac	ammoniac.
Argill	argill.
Oxyd of antimony(A), &c.	antimony(B), &c.

[Note A: The combinations with metallic oxyds were set down by Mr Lavoisier in alphabetical order; their order of affinity being unknown, I have omitted them, as serving no purpose.—E.]

[Note B: All these salts were unknown to the ancient chemists.—A.]

SECT. **XXV**.—*Observations upon Tungstic Acid, and its Combinations.*

Tungstein is a particular metal, the ore of which has frequently been confounded with that of tin. The specific gravity of this ore is to water as 6 to 1; in its form of cristallization it resembles[Pg 252] the garnet, and varies in colour from a pearl-white to yellow and reddish; it is found in several parts of Saxony and Bohemia. The mineral called *Wolfram*, which is frequent in the mines of Cornwall, is likewise an ore of this metal. In all these ores the metal is oxydated; and, in some of them, it appears even to be oxygenated to the state of acid, being combined with lime into a true tungstat of lime.

To obtain the acid free, mix one part of ore of tungstein with four parts of carbonat of potash, and melt the mixture in a crucible, then powder and pour on twelve parts of boiling water, add nitric acid, and the tungstic acid precipitates in a concrete form. Afterwards, to insure the complete oxygenation of the metal, add more nitric acid, and evaporate to dryness, repeating this operation so long as red fumes of nitrous gas are produced. To procure tungstic acid perfectly pure, the fusion of the ore with carbonat of potash must be made in a crucible of platina, otherwise the earth of the common crucibles will mix with the products, and adulterate the acid.[Pg 253]

TABLE *of the Combinations of Tartarous Acid, with the Salifiable Bases, in the Order of Affinity.*

Bases.	Neutral Salts.
Lime	Tartarite of lime

Barytes	barytes.
Magnesia	magnesia.
Potash	potash.
Soda	soda.
Ammoniac	ammoniac.
Argill	argill.
Oxyd of zinc	zinc.
iron	iron.
manganese	manganese.
cobalt	cobalt.
nickel	nickel.
lead	lead.
tin	tin.
copper	copper.
bismuth	bismuth.
antimony	antimony.
arsenic	arsenic.
silver	silver.
mer	mer

	cury		cury.
	gold		gold
	platina		platina

[Pg 254]
SECT. XXVI.—*Observations upon Tartarous Acid, and its Combinations.*

Tartar, or the concretion which fixes to the inside of vessels in which the fermentation of wine is completed, is a well known salt, composed of a peculiar acid, united in considerable excess to potash. Mr Scheele first pointed out the method of obtaining this acid pure. Having observed that it has a greater affinity to lime than to potash, he directs us to proceed in the following manner. Dissolve purified tartar in boiling water, and add a sufficient quantity of lime till the acid be completely saturated. The tartarite of lime which is formed, being almost insoluble in cold water, falls to the bottom, and is separated from the solution of potash by decantation; it is afterwards washed in cold water, and dried; then pour on some sulphuric acid, diluted with eight or nine parts of water, digest for twelve hours in a gentle heat, frequently stirring the mixture; the sulphuric acid combines with the lime, and the tartarous acid is left free. A small quantity of gas, not hitherto examined, is disengaged during this process. At the end of twelve hours, having decanted off the clear liquor, wash the sulphat of lime in cold water, which add to the decanted[Pg 255] liquor, then evaporate the whole, and the tartarous acid is obtained in a concrete form. Two pounds of purified tartar, by means of from eight to ten ounces of sulphuric acid, yield about eleven ounces of tartarous acid.

As the combustible radical exists in excess, or as the acid from tartar is not fully saturated with oxygen, we call it *tartarous acid*, and the neutral salts formed by its combinations with salifiable bases *tartarites*. The base of the tartarous acid is a carbono-hydrous or hydro-carbonous radical, less oxygenated than in the oxalic acid; and it would appear, from the experiments of Mr Hassenfratz, that azote enters into the composition of the tartarous radical, even in considerable quantity. By oxygenating the tartarous acid, it is convertible into oxalic, malic, and acetous acids; but it is probable the proportions of hydrogen and charcoal in the radical are changed during these conversions, and that the difference between these acids does not alone consist in the different degrees of oxygenation.

The tartarous acid is susceptible of two degrees of saturation in its combinations with the fixed alkalies; by one of these a salt is formed with excess of acid, improperly called *cream of tartar*, which in our new nomenclature is named *acidulous tartarite of potash*; by a second or equal degree of saturation a perfectly neutral salt is formed, formerly called *vegetable salt*,[Pg 256] which we name *tartarite of potash*. With soda this acid forms tartarite of soda, formerly called *sal de Seignette*, or *sal polychrest of Rochell.*

SECT. XXVII.—*Observations upon Malic Acid, and its Combinations with the Salifiable Bases*[45].

The malic acid exists ready formed in the sour juice of ripe and unripe apples, and many other fruits, and is obtained as follows: Saturate the juice of apples with potash or soda, and add a proper proportion of acetite of lead dissolved in water; a double decomposition takes place, the malic acid combines with the oxyd of lead and precipitates, being almost insoluble, and the acetite of potash or soda remains in the liquor. The malat of lead being separated by decantation, is washed with cold water, and some dilute sulphuric acid is added; this unites with the lead into an insoluble sulphat, and the malic acid remains free in the liquor.

This acid, which is found mixed with citric and tartarous acid in a great number of fruits, is a kind of medium between oxalic and acetous[Pg 257] acids being more oxygenated than the former, and less so than the latter. From this circumstance, Mr Hermbstadt calls it *imperfect vinegar*, but it differs likewise from acetous acid, by having rather more charcoal, and less hydrogen, in the composition of its radical.

When an acid much diluted has been used in the foregoing process, the liquor contains oxalic as well as malic acid, and probably a little tartarous, these are separated by mixing lime-water with the acids, oxalat, tartarite, and malat of lime are produced; the two former, being insoluble, are precipitated, and the malat of lime remains dissolved; from this the pure malic acid is separated by the acetite of lead, and afterwards by sulphuric acid, as directed above.[Pg 258]

TABLE *of the Combinations of Citric Acid, with the Salifiable Bases, in the Order of Affinity(A).*

Bases.	Neutral Salts.
Barytes	Citrat of barytes.
Lime	lime.
Magnesia	magnesia.
Potash	potash.
Soda	soda.
Ammoniac	ammoniac.
Oxyd of	
zinc	zinc.
manganese	manganese.
iron	iron.
lead	lead.
cobalt	cobalt.
copper	copper.
arsenic	arsenic.
mercury	mercury.

Bases.	Neutral Salts.
antimony	antimony.
silver	silver.
gold	gold.
platina	platina.
Argill	argil.

[Note A: These combinations were unknown to the ancient chemists. The order of affinity of the salifiable bases with this acid was determined by Mr Bergman and by Mr de Breney of the Dijon Academy.—A.][Pg 259]

SECT. XXVIII.—*Observations upon Citric Acid, and its Combinations.*

The citric acid is procured by expression from lemons, and is found in the juices of many other fruits mixed with malic acid. To obtain it pure and concentrated, it is first allowed to depurate from the mucous part of the fruit by long rest in a cool cellar, and is afterwards concentrated by exposing it to the temperature of 4 or 5 degrees below Zero, from 21° to 23° of Fahrenheit, the water is frozen, and the acid remains liquid, reduced to about an eighth part of its original bulk. A lower degree of cold would occasion the acid to be engaged amongst the ice, and render it difficultly separable. This process was pointed out by Mr Georgius.

It is more easily obtained by saturating the lemon-juice with lime, so as to form a citrat of lime, which is insoluble in water; wash this salt, and pour on a proper quantity of sulphuric acid; this forms a sulphat of lime, which precipitates and leaves the citric acid free in the liquor.[Pg 260]

TABLE *of the Combinations of Pyro-lignous Acid with the Salifiable Bases, in the Order of Affinity(A).*

Bases.	Neutral Salts.
Lime	Pyro-mucite of lime.
Barytes	bary tes.
Potash	potash.
Soda	soda.
Magnesia	magnesia.
Ammoniac	ammoniac.
Oxyd of	

zinc	zinc.
manganese	manganese.
iron	iron.
lead	lead.
tin	tin.
cobalt	cobalt.
copper	copper.
nickel	nickel.
arsenic	arsenic.
bismuth	bismuth.
mercury	mercury.
antimony	antimony.
silver	silver.
gold	gold.
platina	platina.
Argill	argil.

[Note A: The above affinities were determined by Messrs de Morveau and EloI Boursier de Clervaux. These combinations were entirely unknown till lately.—A.][Pg 261]

SECT. XXIX.—*Observations upon Pyro-lignous Acid, and its Combinations.*

The ancient chemists observed that most of the woods, especially the more heavy and compact ones, gave out a particular acid spirit, by distillation, in a naked fire; but, before Mr Goetling, who gives an account of his experiments upon this subject in Crell's Chemical Journal for 1779, no one had ever made any inquiry into its nature and properties. This acid appears to be the same, whatever be the wood it is procured from. When first distilled, it is of a brown colour, and considerably impregnated with charcoal and oil; it is purified from these by a second distillation. The pyro-lignous radical is chiefly composed of hydrogen and charcoal.

SECT. XXX.—*Observations upon Pyro-tartarous Acid, and its Combinations with the Salifiable Bases*[46].

The name of *Pyro-tartarous acid* is given to a dilute empyreumatic acid obtained from purified[Pg 262] acidulous tartarite of potash by distillation in a naked fire. To obtain it, let a retort be half filled with powdered tartar, adapt a tubulated recipient, having a bent tube communicating with a bell-glass in a pneumato-chemical apparatus; by gradually raising the fire under the retort, we obtain the pyro-tartarous acid mixed with oil, which is separated by means of a funnel. A vast quantity of carbonic acid gas is disengaged during the distillation. The acid obtained by the above process is much contaminated with oil, which ought to be separated from it. Some authors advise to do this by a second distillation; but the Dijon academicians inform us, that this is attended with great danger from explosions which take place during the process.[Pg 263]

TABLE *of the Combinations of Pyro-mucous Acid, with the Salifiable Bases, in the Order of Affinity(A).*

Bases.	Neutral Salts.
Potash	Pyro-mucite of potash.
Soda	soda.
Barytes	barytes.
Lime	lime.
Magnesia	magnesia.
Ammoniac	ammoniac.
Argill	argil.
Oxyd of zinc	zinc.
manganese	manganese.
iron	iron.
lead	lead.
tin	tin.

Bases	Neutral Salts
cobalt	cobalt.
copper	copper.
nickel	nickel.
arsenic	arsenic.
bismuth	bismuth.
antimony	antimony.

[Note A: All these combinations were unknown to the ancient chemists.—A.][Pg 264]

SECT. XXXI.—*Observations upon Pyro-mucous Acid, and its Combinations.*

This acid is obtained by distillation in a naked fire from sugar, and all the saccharine bodies; and, as these substances swell greatly in the fire, it is necessary to leave seven-eighths of the retort empty. It is of a yellow colour, verging to red, and leaves a mark upon the skin, which will not remove but alongst with the epidermis. It may be procured less coloured, by means of a second distillation, and is concentrated by freezing, as is directed for the citric acid. It is chiefly composed of water and oil slightly oxygenated, and is convertible into oxalic and malic acids by farther oxygenation with the nitric acid.

It has been pretended that a large quantity of gas is disengaged during the distillation of this acid, which is not the case if it be conducted slowly, by means of moderate heat.[Pg 265]

TABLE *of the Combinations of the Oxalic Acid, with the Salifiable Bases, in the Order of Affinity(A).*

Bases.	Neutral Salts.
Lime	Oxalat of lime.
Barytes	barytes.
Magnesia	magnesia.
Potash	potash.
Soda	soda.
Ammoniac	ammoniac.
Argill	argill.

Oxyd of
- zinc.
- iron.
- manganese.
- cobalt.
- nickel.
- lead.
- copper.
- bismuth.
- antimony.
- arsenic.
- mercury.
- silver.
- gold.
- platina.

[Note A: All unknown to the ancient chemists.—A.][Pg 266]

SECT. XXXII.—*Observations upon Oxalic Acid, and its Combinations.*

The oxalic acid is mostly prepared in Switzerland and Germany from the expressed juice of sorrel, from which it cristallizes by being left long at rest; in this state it is partly saturated with potash, forming a true acidulous oxalat of potash, or salt with excess of acid. To obtain it pure, it must be formed artificially by oxygenating sugar, which seems to be the true oxalic radical. Upon one part of sugar pour six or eight parts of nitric acid, and apply a gentle heat; a considerable effervescence takes place, and a great quantity of nitrous gas is disengaged; the nitric acid is decomposed, and its oxygen unites to the sugar: By allowing the liquor to stand at rest, cristals of pure oxalic acid are formed, which must be dried upon blotting paper, to separate any remaining portions of nitric acid; and, to ensure the purity of the acid, dissolve the cristals in distilled water, and cristallize them afresh.

Bases.	Neutral salts.	Names of the resulting neutral salts according to the old nomenclature.
Barytes	Acetite of barytes	Unknown to the ancients. Discovered by Mr de Morveau, who calls it *barotic acéte*.
Potash	potash	Secret terra foliata tartari of Muller. Arcanum tartari of Basil Valentin and Paracelsus. Purgative magistery of tartar of Schroëder. Essential salt of wine of Zwelfer. Regenerated tartar of Tachenius. Diuretic salt of Sylvius and Wilson.
Soda	soda	Foliated earth with base of mineral alkali. Mineral or crystallisable foliated earth. Mineral acetous salt.
Lime	lime	Salt of chalk, coral, or crabs eyes; mentioned by Hartman.
Magnesia	magnesia	First mentioned by Mr Wenzel.
Ammoniac	ammoniac	Spiritus Mindereri. Ammoniacal acetous salt.
Oxyd of zinc	zinc	Known to Glauber, Schwedemberg, Respour, Pott, de Lassone, and Wenzel, but not named.
manganese	manganese	Unknown to the ancients.
iron	iron	Martial vinegar. Described by Monnet, Wenzel, and the Duke d'Ayen.
lead	lead	Sugar, vinegar, and salt of lead or Saturn.
tin	tin	Known to Lemery, Margraff, Monnet, Weslendorf, and Wenzel, but not named.
cobalt	cobalt	Sympathetic ink of Mr Cadet.
copper	copper	Verdigris, crystals of verditer, verditer, distilled verdigris, crystals of Venus or of copper.
nickel	nickel	Unknown to the ancients.
arsenic	arsenic	Arsenico-acetous fuming liquor, liquid phosphorus of Mr Cadet.
bismuth	bismuth	Sugar of bismuth of Mr Geoffroi. Known to Gellert, Pott, Weslendorf, Bergman, and de Morveau.

mercury	mercury	Mercurial foliated earth, Keyser's famous antivenereal remedy. Mentioned by Gebaver in 1748; known to Helot, Margraff, Baumé, Bergman, and de Morveau.
antimony	antimony	Unknown.
silver	silver	Described by Margraff, Monnet, and Wenzel; unknown to the ancients.
gold	gold	Little known, mentioned by Schroëder and Juncker.
platina	platina	Unknown.
Argil l	argill	According to Mr Wenzel, vinegar dissolves only a very small proportion of argill.

[Pg 267]
From the liquor remaining after the first cristallization of the oxalic acid we may obtain malic acid by refrigeration: This acid is more oxygenated than the oxalic; and, by a further oxygenation, the sugar is convertible into acetous acid, or vinegar.

The oxalic acid, combined with a small quantity of soda or potash, has the property, like the tartarous acid, of entering into a number of combinations without suffering decomposition: These combinations form triple salts, or neutral salts with double bases, which ought to have proper names. The salt of sorrel, which is potash having oxalic acid combined in excess, is named acidulous oxalat of potash in our new nomenclature.

The acid procured from sorrel has been known to chemists for more than a century, being mentioned by Mr Duclos in the Memoirs of the Academy for 1688, and was pretty accurately described by Boerhaave; but Mr Scheele first showed that it contained potash, and demonstrated its identity with the acid formed by the oxygenation of sugar.

SECT. XXXIII.—*Observations upon Acetous Acid, and its Combinations.*

This acid is composed of charcoal and hydrogen united together, and brought to the state of an acid by the addition of oxygen; it is consequently formed by the same elements with[Pg 268] the tartarous oxalic, citric, malic acids, and others, but the elements exist in different proportions in each of these; and it would appear that the acetous acid is in a higher state of oxygenation than these other acids. I have some reason to believe that the acetous radical contains a small portion of azote; and, as this element is not contained in the radicals of any vegetable acid except the tartarous, this circumstance is one of the causes of difference. The acetous acid, or vinegar, is produced by exposing wine to a gentle heat, with the addition of some ferment: This is usually the ley, or mother, which has separated from other vinegar during fermentation, or some similar matter. The spiritous part of the wine, which consists of charcoal and hydrogen, is oxygenated, and converted into vinegar: This operation can only take place with free access of air, and is always attended by a diminution of the air employed in consequence of the absorption of oxygen; wherefore, it ought always to be carried on in vessels only half filled with the vinous liquor submitted to the acetous fermentation. The acid formed during this process is very volatile, is mixed with a large proportion of water, and with many foreign substances; and, to obtain it pure, it is distilled in stone or glass vessels by a gentle fire. The acid which passes over in distillation is somewhat changed by the[Pg 269] process, and is not exactly of the same nature with what remains in the alembic, but seems less oxygenated: This circumstance has not been formerly observed by chemists.

Distillation is not sufficient for depriving this acid of all its unnecessary water; and, for this purpose, the best way is by exposing it to a degree of cold from 4° to 6° below the freezing point, from 19° to 23° of Fahrenheit; by this means the aqueous part becomes frozen, and leaves the acid in a liquid state, and considerably concentrated. In the usual temperature of the air, this acid can only exist in the gasseous form, and can only be retained by combination with a large proportion of water. There are other chemical processes for obtaining the acetous acid, which consist in oxygenating the tartarous, oxalic, or malic acids, by means of nitric acid; but there is reason to believe the proportions of the elements of the radical are changed during this process. Mr Hassenfratz is at present engaged in repeating the experiments by which these conversions are said to be produced.

The combinations of acetous acid with the various salifiable bases are very readily formed; but most of the resulting neutral salts are not cristallizable, whereas those produced by the tartarous and oxalic acids are, in general, hardly soluble. Tartarite and oxalat of lime are[Pg 270] not soluble in any sensible degree: The malats are a medium between the oxalats and acetites, with respect to solubility, and the malic acid is in the middle degree of saturation between the oxalic and acetous acids. With this, as with all the acids, the metals require to be oxydated previous to solution.

The ancient chemists knew hardly any of the salts formed by the combinations of acetous acid with the salifiable bases, except the acetites of potash, soda, ammoniac, copper, and lead. Mr Cadet discovered the acetite of arsenic[47]; Mr Wenzel, the Dijon academicians Mr de Lassone, and Mr Proust, made us acquainted with the properties of the other acetites. From the property which acetite of potash possesses, of giving out ammoniac in distillation, there is some reason to suppose, that, besides charcoal and hydrogen, the acetous radical contains a small proportion of azote, though it is not impossible but the above production of ammoniac may be occasioned by the decomposition of the potash.

[Pg 271]

TABLE of the *Combinations of Acetic Acid with the Salifiable Bases, in the order of affinity.*

Bases.	Neutral Salts.
Barytes	Acetat of barytes.
Potash	potash.
Soda	soda.
Lime	lime.
Magnesia	magnesia.
Ammoniac	ammoniac.
Oxyd of zinc	zinc.
manganese	manganese.
iron	iron
lea	lead

Bases	Neutral Salts
...	...
tin	tin.
cobalt	cobalt.
copper	copper.
nickel	nickel.
arsenic	arsenic.
bismuth	bismuth.
mercury	mercury.
antimony	antimony.
silver	silver.
gold	gold.
platina	platina.
Argill	argil.

Note.—All these salts were unknown to the ancients; and even those chemists who are most versant in modern discoveries, are yet at a lose whether the greater part of the salts produced by the oxygenated acetic radical belong properly to the class of acetites, or to that of acetats.—A.[Pg 272]

SECT. XXXIV.—*Observations upon Acetic Acid, and its Combinations.*

We have given to radical vinegar the name of acetic acid, from supposing that it consists of the same radical with that of the acetous acid, but more highly saturated with oxygen. According to this idea, acetic acid is the highest degree of oxygenation of which the hydro-carbonous radical is susceptible; but, although this circumstance be extremely probable, it requires to be confirmed by farther, and more decisive experiments, before it be adopted as an absolute chemical truth. We procure this acid as follows: Upon three parts acetite of potash or of copper, pour one part of concentrated sulphuric acid, and, by distillation, a very highly concentrated vinegar is obtained, which we call acetic acid, formerly named radical vinegar. It is not hitherto rigorously proved that this acid is more highly oxygenated than the acetous acid, nor that the difference between them may not consist in a different proportion between the elements of the radical or base.[Pg 273]

TABLE *of the Combinations of Succinic Acid with the Salifiable Bases, in the order of Affinity.*

Bases.	Neutral Salts.

Barytes		Succinat of	barytes.
Lime			lime.
Potash			potash.
Soda			soda.
Ammoniac			ammoniac.
Magnesia			magnesia.
Argill			argil.
Oxyd	of zinc		zinc.
	iron		iron.
	manganese		manganese.
	cobalt		cobalt.
	nickel		nickel.
	lead		lead.
	tin		tin.
	copper		copper.
	bismuth		bismuth.
	antimony		antimony.
	arsenic		arsenic.
	mercury		mercury.
	silv		silve

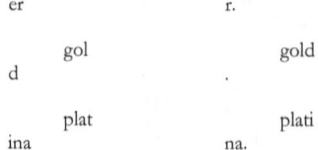

Note.—All the succinats were unknown to the ancient chemists.—A.[Pg 274]

SECT. XXXV.—*Observations upon Succinic Acid, and its Combinations.*

The succinic acid is drawn from amber by sublimation in a gentle heat, and rises in a concrete form into the neck of the subliming vessel. The operation must not be pushed too far, or by too strong a fire, otherwise the oil of the amber rises alongst with the acid. The salt is dried upon blotting paper, and purified by repeated solution and crystallization.

This acid is soluble in twenty-four times its weight of cold water, and in a much smaller quantity of hot water. It possesses the qualities of an acid in a very small degree, and only affects the blue vegetable colours very slightly. The affinities of this acid, with the salifiable bases, are taken from Mr de Morveau, who is the first chemist that has endeavoured to ascertain them.[Pg 275]

SECT. XXXVI.—*Observations upon Benzoic Acid, and its Combinations with Salifiable Bases*[48].

This acid was known to the ancient chemists under the name of Flowers of Benjamin, or of Benzoin, and was procured, by sublimation, from the gum or resin called Benzoin: The means of procuring it, *via humida*, was discovered by Mr Geoffroy, and perfected by Mr Scheele. Upon benzoin, reduced to powder, pour strong lime-water, having rather an excess of lime; keep the mixture continually stirring, and, after half an hour's digestion, pour off the liquor, and use fresh portions of lime-water in the same manner, so long as there is any appearance of neutralization. Join all the decanted liquors, and evaporate, as far as possible, without occasioning cristallization, and, when the liquor is cold, drop in muriatic acid till no more precipitate is formed. By the former part of the process a benzoat of lime is formed, and, by the latter, the muriatic acid combines with the lime, forming muriat of lime, which remains[Pg 276]dissolved, while the benzoic acid, being insoluble, precipitates in a concrete state.

SECT. XXXVII.—*Observations upon Camphoric Acid, and its Combinations with Salifiable Bases*[49].

Camphor is a concrete essential oil, obtained, by sublimation, from a species of laurus which grows in China and Japan. By distilling nitric acid eight times from camphor, Mr Kosegarten converted it into an acid analogous to the oxalic; but, as it differs from that acid in some circumstances, we have thought necessary to give it a particular name, till its nature be more completely ascertained by farther experiment.

As camphor is a carbono-hydrous or hydro-carbonous radical, it is easily conceived, that, by oxygenation, it should form oxalic, malic, and several other vegetable acids: This conjecture is rendered not improbable by the experiments of Mr Kosegarten; and the principal phenomena exhibited in the combinations of camphoric acid with the salifiable bases, being[Pg 277] very similar to those of the oxalic and malic acids, lead me to believe that it consists of a mixture of these two acids.

SECT. XXXVIII.—*Observations upon Gallic Acid, and its Combinations with Salifiable Bases*[50].

The Gallic acid, formerly called Principle of Astringency, is obtained from gall nuts, either by infusion or decoction with water, or by distillation with a very gentle heat. This acid has only been attended to within these few years. The Committee of the Dijon Academy have followed it through all its combinations, and give the best account of it hitherto produced. Its acid properties are very weak; it reddens the tincture of turnsol, decomposes sulphurets, and unites to all the metals when they have been previously dissolved in some other acid. Iron, by this combination, is precipitated of a very deep blue or violet colour. The radical of this acid, if it deserves the name of one, is hitherto entirely unknown; it is contained in[Pg 278] oak willow, marsh iris, the strawberry, nymphea, Peruvian bark, the flowers and bark of pomgranate, and in many other woods and barks.

SECT. XXXIX.—*Observations upon Lactic Acid, and its Combinations with Salifiable Bases*[51].

The only accurate knowledge we have of this acid is from the works of Mr Scheele. It is contained in whey, united to a small quantity of earth, and is obtained as follows: Reduce whey to one eighth part of its bulk by evaporation, and filtrate, to separate all its cheesy matter; then add as much lime as is necessary to combine with the acid; the lime is afterwards disengaged by the addition of oxalic acid, which combines with it into an insoluble neutral salt. When the oxalat of lime has been separated by decantation, evaporate the remaining liquor to the consistence of honey; the lactic acid is dissolved by alkohol, which does not unite with the sugar of milk and other foreign matters;[Pg 279] these are separated by filtration from the alkohol and acid; and the alkohol being evaporated, or distilled off, leaves the lactic acid behind.

This acid unites with all the salifiable bases forming salts which do not cristallize; and it seems considerably to resemble the acetous acid.[Pg 280]

TABLE *of the Combinations of Saccholactic Acid with the Salifiable Bases, in the Order of Affinity.*

Bases.	Neutral Salts.
Lime	Saccholat of lime.
Barytes	barytes.
Magnesia	magnesia.
Potash	potash.
Soda	soda.
Ammoniac	ammoniac.
Argill	argill.
Oxyd of zinc	zinc.
manganese	manganese.
iron	iron.
lead	lead.
tin	tin.
cob	cob

alt	alt.
copper	copper.
nickel	nickel.
arsenic	arsenic.
bismuth	bismuth.
mercury	mercury.
antimony	antimony.
silver	silver.

Note.—All these were unknown to the ancient chemists.—A.[Pg 281]

SECT. XL.—*Observations upon Saccholactic Acid, and its Combinations.*

A species of sugar may be extracted, by evaporation, from whey, which has long been known in pharmacy, and which has a considerable resemblance to that procured from sugar canes. This saccharine matter, like ordinary sugar, may be oxygenated by means of nitric acid: For this purpose, several portions of nitric acid are distilled from it; the remaining liquid is evaporated, and set to cristallize, by which means cristals of oxalic acid are procured; at the same time a very fine white powder precipitates, which is the saccholactic acid discovered by Scheele. It is susceptible of combining with the alkalies, ammoniac, the earths, and even with the metals: Its action upon the latter is hitherto but little known, except that, with them, it forms difficultly soluble salts. The order of affinity in the table is taken from Bergman.[Pg 282]

TABLE *of the Combinations of Formic Acid, with the Salifiable Bases, in the Order of Affinity.*

Bases.	Neutral Salts.
Barytes	Formiat of barytes.
Potash	potash.
Soda	soda.
Lime	lime.
Magnesia	magnesia.
Ammoniac	ammoniac.

Oxyd of	
zinc	zinc.
manganese	manganese.
iron	iron.
lead	lead.
tin	tin.
cobalt	cobalt.
copper	copper.
nickel	nickel.
bismuth	bismuth.
silver	silver.
Argill	argil.

Note.—All unknown to the ancient chemists.—A.[Pg 283]

SECT. XLI.—*Observations upon Formic Acid, and its Combinations.*

This acid was first obtained by distillation from ants, in the last century, by Samuel Fisher. The subject was treated of by Margraff in 1749, and by Messrs Ardwisson and Ochrn of Leipsic in 1777. The formic acid is drawn from a large species of red ants, *formica rufa, Lin.* which form large ant hills in woody places. It is procured, either by distilling the ants with a gentle heat in a glass retort or an alembic; or, after having washed the ants in cold water, and dried them upon a cloth, by pouring on boiling water, which dissolves the acid; or the acid may be procured by gentle expression from the insects, in which case it is stronger than in any of the former ways. To obtain it pure, we must rectify, by means of distillation, which separates it from the uncombined oily and charry matter; and it may be concentrated by freezing, in the manner recommended for treating the acetous acid.[Pg 284]

SECT. XLII.—*Observations upon Bombic Acid, and its Combinations with Acidifiable Bases*[52].

The juices of the silk worm seem to assume an acid quality when that insect changes from a larva to a chrysalis. At the moment of its escape from the latter to the butterfly form, it emits a reddish liquor which reddens blue paper, and which was first attentively observed by Mr Chaussier of the Dijon academy, who obtains the acid by infusing silk worm chrysalids in alkohol, which dissolves their acid without being charged with any of the gummy parts of the insect; and, by evaporating the alkohol, the acid remains tollerably pure. The properties and affinities of this acid are not hitherto ascertained with any precision; and we have reason to believe that analogous acids may be procured from other insects. The radical of this acid is probably, like that of the other acids from the animal kingdom, composed of charcoal, hydrogen, and azote, with the addition, perhaps, of phosphorus.

[Pg 285]

TABLE *of the Combinations of Sebacic Acid, with the Salifiable Bases, in the Order of Affinity.*

Bases.	Neutral Salts.
Barytes	Sebat of barytes.
Potash	potash.
Soda	soda.
Lime	lime.
Magnesia	magnesia.
Ammoniac	ammoniac.
Argill	argill.
Oxyd of zinc	zinc.
manganese	manganese.
iron	iron.
lead	lead.
tin	tin.
cobalt	cobalt.
copper	copper.
nickel	nickel.
arsenic	arsenic.

bismuth	bismuth.
mercury	mercury.
antimony	antimony.
silver	silver.

Note.—All these were unknown to the ancient chemists.—A.[Pg 286]

SECT. XLIII.—*Observations upon Sebacid Acid, and its Combinations.*

To obtain the sebacic acid, let some suet be melted in a skillet over the fire, alongst with some quick-lime in fine powder, and constantly stirred, raising the fire towards the end of the operation, and taking care to avoid the vapours, which are very offensive. By this process the sebacic acid unites with the lime into a sebat of lime, which is difficultly soluble in water; it is, however, separated from the fatty matters with which it is mixed by solution in a large quantity of boiling water. From this the neutral salt is separated by evaporation; and, to render it pure, is calcined, redissolved, and again cristallized. After this we pour on a proper quantity of sulphuric acid, and the sebacic acid passes over by distillation.[Pg 287]

SECT. XLIV.—*Observations upon the Lithic Acid, and its Combinations with the Salifiable Bases*[53].

From the later experiments of Bergman and Scheele, the urinary calculus appears to be a species of salt with an earthy basis; it is slightly acidulous, and requires a large quantity of water for solution, three grains being scarcely soluble in a thousand grains of boiling water, and the greater part again cristallizes when cold. To this concrete acid, which Mr de Morveau calls Lithiasic Acid, we give the name of Lithic Acid, the nature and properties of which are hitherto very little known. There is some appearance that it is an acidulous neutral salt, or acid combined in excess with a salifiable base; and I have reason to believe that it really is an acidulous phosphat of lime; if so, it must be excluded from the class of peculiar acids.

[Pg 288]

TABLE *of the Combinations of the Prussic Acid with the Salifiable Bases, in the order of affinity.*

Bases.	Neutral Salts.
Potash	Prussiat of potash.
Soda	sod a.
Ammoniac	ammoniac.
Lime	lime.
Barytes	barytes.
Magnesia	magnesia.

Oxyd	of zinc	zinc.
	iron	iron.
	manganese	manganese.
	cobalt	cobalt.
	nickel	nickel.
	lead	lead.
	tin	tin.
	copper	copper.
	bismuth	bismuth.
	antimony	antimony.
	arsenic	arsenic.
	silver	silver.
	mercury	mercury.
	gold	gold.
	platina	platina.

Note.—All these were unknown to former chemists.—A.
[Pg 289]

Observations upon the Prussic Acid, and its Combinations.

As the experiments which have been made hitherto upon this acid seem still to leave a considerable degree of uncertainty with regard to its nature, I shall not enlarge upon its properties, and the means of procuring it pure and dissengaged from combination. It combines with iron, to which it communicates a blue colour, and is equally susceptible of entering into combination with most of the other metals, which are precipitated from it by the alkalies, ammoniac, and lime, in consequence of greater affinity. The Prussic radical, from the experiments of Scheele, and especially from those of Mr Berthollet, seems composed of charcoal and azote; hence it is an acid with a double base. The phosphorus which has been found combined with it appears, from the experiments of Mr Hassenfratz, to be only accidental.

Although this acid combines with alkalies, earths, and metals, in the same way with other acids, it possesses only some of the properties we have been in use to attribute to acids, and it may consequently be improperly ranked here in[Pg 290] the class of acids; but, as I have already observed, it is difficult to form a decided opinion upon the nature of this substance until the subject has been farther elucidated by a greater number of experiments.

FOOTNOTES:

[36]See Memoirs of the Academy for 1776, p. 671. and for 1778, p. 535,—A.

[37]See Part I. Chap. XI. upon this subject.—A.

[38]See Part I. Chap. XI. upon the application of these names according to the proportions of the two ingredients.—A

[39]See Part I. Chap. XII. upon this subject.—A.

[40]Those who wish to see what has been said upon this great chemical question by Messrs de Morveau, Berthollet, De Fourcroy, and myself, may consult our translation of Mr Kirwan's Essay upon Phlogiston.—A.

[41]Saltpetre is likewise procured in large quantities by lixiviating the natural soil in some parts of Bengal, and of the Russian Ukrain.—E.

[42]Commonly called *Derbyshire spars*.—E.

[43]I have not added the Table of these combinations, as the order of their affinity is entirely unknown; they are called *molybdats of argil, antimony,potash*, &c.—E.

[44]This acid was discovered by Mr Scheele, to whom chemistry is indebted for the discovery of several other acids.—A.

[45]I have omitted the Table, as the order of affinity is unknown, and is given by Mr Lavoisier only in alphabetical order. All the combinations of malic acid with salifiable bases, which are named *malats*, were unknown to the ancient chemists.—E.

[46]The order of affinity of the salifiable bases with this acid is hitherto unknown. Mr Lavoisier, from its similarity to pyro-lignous acid, supposes the order to be the same in both; but, as this is not ascertained by experiment, the table is omitted. All these combinations, called *Pyro-tartarites*, were unknown till lately—E.

[47]Savans Etrangers, Vol. III.

[48]These combinations are called Benzoats of Lime, Potash, Zinc, &c.; but, as the order of affinity is unknown, the alphabetical table is omitted, as unnecessary.—E.

[49]These combinations, which were all unknown to the ancients, are called Camphorats. The table is omitted, as being only in alphabetical order.—E.

[50]These combinations, which are called Gallats, were all unknown to the ancients; and the order of their affinity is not hitherto established.—A.

[51]These combinations are called Lactats; they were all unknown to the ancient chemists, and their affinities have not yet been ascertained.—A.

[52]These combinations named Bombats were unknown to the ancient chemists; and the affinities of the salifiable bases with the bombic acid are hitherto undetermined.—A.

[53]All the combinations of this acid, should it finally turn out to be one, were unknown to the ancient chemists, and its affinities with the salifiable bases have not been hitherto determined.—A.

[Pg 291]

PART III.
Description of the Instruments and Operations of Chemistry.

INTRODUCTION.

In the two former parts of this work I designedly avoided being particular in describing the manual operations of chemistry, because I had found from experience, that, in a work appropriated to reasoning, minute descriptions of processes and of plates interrupt the chain of ideas, and render the attention necessary both difficult and tedious to the reader. On the other hand, if I had confined myself to the summary descriptions hitherto given, beginners could have only acquired very vague conceptions of practical chemistry from my work, and must have wanted both confidence and interest in operations they could neither repeat nor[Pg 292] thoroughly comprehend. This want could not have been supplied from books; for, besides that there are not any which describe the modern instruments and experiments sufficiently at large, any work that could have been consulted would have presented these things under a very different order of arrangement, and in a different chemical language, which must greatly tend to injure the main object of my performance.

Influenced by these motives, I determined to reserve, for a third part of my work, a summary description of all the instruments and manipulations relative to elementary chemistry. I considered it as better placed at the end, rather than at the beginning of the book, because I must have been obliged to suppose the reader acquainted with circumstances which a beginner cannot know, and must therefore have read the elementary part to become acquainted with. The whole of this third part may therefore be considered as resembling the explanations of plates which are usually placed at the end of academic memoirs, that they may not interrupt the connection of the text by lengthened description. Though I have taken great pains to render this part clear and methodical, and have not omitted any essential instrument or apparatus, I am far from pretending by it to set aside the necessity of attendance upon lectures and laboratories,[Pg 293]for such as wish to acquire accurate knowledge of the science of chemistry. These should familiarise themselves to the employment of apparatus, and to the performance of experiments by actual experience. *Nihil est in intellectu quod non prius fuerit in sensu*, the motto which the celebrated Rouelle caused to be painted in large characters in a conspicuous part of his laboratory, is an important truth never to be lost sight of either by teachers or students of chemistry.

Chemical operations may be naturally divided into several classes, according to the purposes they are intended for performing. Some may be considered as purely mechanical, such as the determination of the weight and bulk of bodies, trituration, levigation, searching, washing, filtration, &c. Others may be considered as real chemical operations, because they are performed by means of chemical powers and agents; such are solution, fusion, &c. Some of these are intended for separating the elements of bodies from each other, some for reuniting these elements together; and some, as combustion, produce both these effects during the same process.

Without rigorously endeavouring to follow the above method, I mean to give a detail of the chemical operations in such order of arrangement as seemed best calculated for conveying[Pg 294] instruction. I shall be more particular in describing the apparatus connected with modern chemistry, because these are hitherto little known by men who have devoted much of their time to chemistry, and even by many professors of the science.

[Pg 295]

CHAP. I.
Of the Instruments necessary for determining the Absolute and Specific Gravities of Solid and Liquid Bodies.

The best method hitherto known for determining the quantities of substances submitted to chemical experiment, or resulting from them, is by means of an accurately constructed beam and scales, with properly regulated weights, which well known operation is called *weighing*. The denomination and quantity of the weights used as an unit or standard for this purpose are extremely arbitrary, and vary not only in different kingdoms, but even in different provinces of the same kingdom, and in different cities of the same province. This variation is of infinite consequence to be well understood in commerce and in the arts; but, in chemistry, it is of no moment what particular denomination of weight be employed, provided the results of experiments be expressed in convenient fractions of the same denomination. For this purpose, until all the weights used in society be reduced to the same standard, it will be sufficient for chemists in different parts to use the common[Pg 296] pound of their own country as the unit or standard, and to express all its fractional parts in decimals, instead of the arbitrary divisions now in use. By this means the chemists of all countries will be thoroughly understood by each other, as, although the absolute weights of the ingredients and products cannot be known, they will readily, and without calculation, be able to determine the relative proportions of these to each other with the utmost accuracy; so that in this way we shall be possessed of an universal language for this part of chemistry.

With this view I have long projected to have the pound divided into decimal fractions, and I have of late succeeded through the assistance of Mr Fourche balance-maker at Paris, who has executed it for me with great accuracy and judgment. I recommend to all who carry on experiments to procure similar divisions of the pound, which they will find both easy and simple in its application, with a very small knowledge of decimal fractions[54].

[Pg 297]

As the usefulness and accuracy of chemistry depends entirely upon the determination of the weights of the ingredients and products both before and after experiments, too much precision cannot be employed in this part of the subject; and, for this purpose, we must be provided with good instruments. As we are often obliged, in chemical processes, to ascertain, within a grain or less, the tare or weight of large and heavy instruments, we must have beams made with peculiar niceness by accurate workmen, and these must always be kept apart from the laboratory in some place where the vapours of acids, or other corrosive liquors, cannot have access, otherwise the steel will rust, and the accuracy of the balance be destroyed. I have three sets, of different sizes, made by Mr Fontin with the utmost nicety, and, excepting those made by Mr Ramsden of London, I do not think any can compare with them for precision and sensibility. The largest of these is about three feet long in the beam for large weights, up to fifteen or twenty pounds; the second, for weights of eighteen or twenty ounces, is exact to a tenth part of a grain; and the smallest, calculated only for weighing about one gros, is sensibly affected by the five hundredth part of a grain.

Besides these nicer balances, which are only used for experiments of research, we must have[Pg 298] others of less value for the ordinary purposes of the laboratory. A large iron balance, capable of weighing forty or fifty pounds within half a dram, one of a middle size, which may ascertain eight or ten pounds, within ten or twelve grains, and a small one, by which about a pound may be determined, within one grain.

We must likewise be provided with weights divided into their several fractions, both vulgar and decimal, with the utmost nicety, and verified by means of repeated and accurate trials in the nicest scales; and it requires some experience, and to be accurately acquainted with the different weights, to be able to use them properly. The best way of precisely ascertaining the weight of any particular substance is to weigh it twice, once with the decimal divisions of the pound, and another time with the common subdivisions or vulgar fractions, and, by comparing these, we attain the utmost accuracy.

By the specific gravity of any substance is understood the quotient of its absolute weight divided by its magnitude, or, what is the same, the weight of a determinate bulk of any body. The weight of a determinate magnitude of water has been generally assumed as unity for this purpose; and we express the specific gravity of gold, sulphuric acid, &c. by saying, that gold is nineteen times, and sulphuric acid twice the weight of water, and so of other bodies.[Pg 299]

It is the more convenient to assume water as unity in specific gravities, that those substances whose specific gravity we wish to determine, are most commonly weighed in water for that purpose. Thus, if we wish to determine the specific gravity of gold flattened under the hammer, and supposing the piece of gold to weigh *8 oz. 4 gros 2-1/2 grs.* in the air[55], it is suspended by means of a fine metallic wire under the scale of a hydrostatic balance, so as to be entirely immersed in water, and again weighed. The piece of gold in Mr Brisson's experiment lost by this means *3 gros 37 grs.*; and, as it is evident that the weight lost by a body weighed in water is precisely equal to the weight of the water displaced, or to that of an equal volume of water, we may conclude, that, in equal magnitudes, gold weighs *4893-1/2 grs.* and water *253 grs.*which, reduced to unity, gives 1.0000 as the specific gravity of water, and 19.3617 for that of gold. We may operate in the same manner with all solid substances. We have rarely any occasion, in chemistry, to determine the specific gravity of solid bodies, unless when operating upon alloys or metallic glasses; but we have very frequent necessity to ascertain that of fluids, as it is often the only means of judging of their purity or degree of concentration.

[Pg 300]

This object may be very fully accomplished with the hydrostatic balance, by weighing a solid body; such, for example, as a little ball of rock cristal suspended by a very fine gold wire, first in the air, and afterwards in the fluid whose specific gravity we wish to discover. The weight lost by the cristal, when weighed in the liquor, is equal to that of an equal bulk of the liquid. By repeating this operation successively in water and different fluids, we can very readily ascertain, by a simple and easy calculation, the relative specific gravities of these fluids, either with respect to each other or to water. This method is not, however, sufficiently exact, or, at least, is rather troublesome, from its extreme delicacy, when used for liquids differing but little in specific gravity from water; such, for instance, as mineral waters, or any other water containing very small portions of salt in solution.

In some operations of this nature, which have not hitherto been made public, I employed an instrument of great sensibility for this purpose with great advantage. It consists of a hollow cylinder, *A b c f*, Pl. vii. fig. 6. of brass, or rather of silver, loaded at its bottom, b c f, with tin, as represented swimming in a jug of water, *l m n o*. To the upper part of the cylinder is attached a stalk of silver wire, not more than three fourths of a line diameter, surmounted by[Pg 301] a little cup *d*, intended for containing weights; upon the stalk a mark is made at *g*, the use of which we shall presently explain. This cylinder may be made of any size; but, to be accurate, ought at least to displace four pounds of water. The weight of tin with which this instrument is loaded ought to be such as will make it remain almost in equilibrium in distilled water, and should not require more than half a dram, or a dram at most, to make it sink to *g*.

We must first determine, with great precision, the exact weight of the instrument, and the number of additional grains requisite for making it sink, in distilled water of a determinate temperature, to the mark: We then perform the same experiment upon all the fluids of which we wish to ascertain the specific gravity, and, by means of calculation, reduce the observed differences to a common standard of cubic feet, pints or pounds, or of decimal fractions, comparing them with water. This method, joined to experiments with certain reagents[56], is one of the best for determining the quality of waters, and is even capable of pointing out differences which escape the most accurate chemical analysis. I shall, at some future[Pg 302] period, give an account of a very extensive set of experiments which I have made upon this subject.

These metallic hydrometers are only to be used for determining the specific gravities of such waters as contain only neutral salts or alkaline substances; and they may be constructed with different degrees of ballast for alkohol and other spiritous liquors. When the specific gravities of acid liquors are to be ascertained, we must use a glass hydrometer, as represented Pl. vii. fig. 14[57]. This consists of a hollow cylinder of glass, *a b c f*, hermetically sealed at its lower end, and drawn out at the upper into a capillary tube *a*, ending in the little cup or bason *d*. This instrument is ballasted with more or less mercury, at the bottom of the cylinder introduced through the tube, in proportion to the weight of the liquor intended to be examined: We may introduce a small graduated slip of paper into the tube *a d*; and, though these degrees do not exactly correspond to the fractions of grains in the different liquors, they may be rendered very useful in calculation.

What is said in this chapter may suffice, without farther enlargement, for indicating the[Pg 303] means of ascertaining the absolute and specific gravities of solids and fluids, as the necessary instruments are generally known, and may easily be procured: But, as the instruments I have used for measuring the gasses are not any where described, I shall give a more detailed account of these in the following chapter.

FOOTNOTES:

[54]Mr Lavoisier gives, in this part of his work, very accurate directions for reducing the common subdivisions of the French pound into decimal fractions, and *vice versa*, by means of tables subjoined to this 3d part. As these instructions, and the table, would be useless to the British chemist, from the difference between the subdivisions of the French and Troy pounds, I have omitted them, but have subjoined in the appendix accurate rules for converting the one into the other.—E.

[55]Vide Mr Brisson's Essay upon Specific Gravity, p. 5.—A.

[56]For the use of these reagents see Bergman's excellent treatise upon the analysis of mineral waters, in his Chemical and Physical Essays.—E.

[57]Three or four years ago, I have seen similar glass hydrometers, made for Dr Black by B. Knie, a very ingenious artist of this city.—E.

[Pg 304]

CHAP. II.
Of Gazometry, or the Measurement of the Weight and Volume of Aëriform Substances.

SECT. I.
Description of the Pneumato-chemical Apparatus.

The French chemists have of late applied the name of *pneumato-chemical apparatus* to the very simple and ingenious contrivance, invented by Dr Priestley, which is now indispensibly necessary to every laboratory. This consists of a wooden trough, of larger or smaller dimensions as is thought convenient, lined with plate-lead or tinned copper, as represented in perspective, Pl. V. In Fig. 1. the same trough or cistern is supposed to have two of its sides cut away, to show its interior construction more distinctly. In this apparatus, we distinguish between the shelf ABCD Fig. 1. and 2. and the bottom or body of the cistern FGHI Fig. 2.[Pg 305] The jars or bell-glasses are filled with water in this deep part, and, being turned with their mouths downwards, are afterwards set upon the shelf ABCD, as shown Plate X. Fig. 1. F. The upper parts of the sides of the cistern above the level of the shelf are called the *rim* or *borders*.

The cistern ought to be filled with water, so as to stand at least an inch and a half deep upon the shelf, and it should be of such dimensions as to admit of at least one foot of water in every direction in the well. This size is sufficient for ordinary occasions; but it is often convenient, and even necessary, to have more room; I would therefore advise such as intend to employ themselves usefully in chemical experiments, to have this apparatus made of considerable magnitude, where their place of operating will allow. The well of my principal cistern holds four cubical feet of water, and its shelf has a surface of fourteen square feet; yet, in spite of this size, which I at first thought immoderate, I am often straitened for room.

In laboratories, where a considerable number of experiments are performed, it is necessary to have several lesser cisterns, besides the large one, which may be called the *general magazine*; and even some portable ones, which may be moved when necessary, near a furnace, or wherever they may be wanted. There are likewise some operations which dirty the water of the apparatus,[Pg 306] and therefore require to be carried on in cisterns by themselves.

It were doubtless considerably cheaper to use cisterns, or iron-bound tubs, of wood simply dove-tailed, instead of being lined with lead or copper; and in my first experiments I used them made in that way; but I soon discovered their inconvenience. If the water be not always kept at the same level, such of the dovetails as are left dry shrink, and, when more water is added, it escapes through the joints, and runs out.

We employ cristal jars or bell glasses, Pl. V. Fig. 9. A. for containing the gasses in this apparatus; and, for transporting these, when full of gas, from one cistern to another, or for keeping them in reserve when the cistern is too full, we make use of a flat dish BC, surrounded by a standing up rim or border, with two handles DE for carrying it by.

After several trials of different materials, I have found marble the best substance for constructing the mercurial pneumato-chemical apparatus, as it is perfectly impenetrable by mercury, and is not liable, like wood, to separate at the junctures, or to allow the mercury to escape through chinks; neither does it run the risk of breaking, like glass, stone-ware, or porcelain. Take a block of marble BCDE, Plate V. Fig. 3. and 4. about two feet long, 15 or 18 inches[Pg 307] broad, and ten inches thick, and cause it to be hollowed out as at *m n* Fig. 5. about four inches deep, as a reservoir for the mercury; and, to be able more conveniently to fill the jars, cut the gutter T V, Fig. 3. 4. and 5. at least four inches deeper; and, as this trench may sometimes prove troublesome, it is made capable of being covered at pleasure by thin boards, which slip into the grooves *x y*, Fig. 5. I have two marble cisterns upon this construction, of different sizes, by which I can always employ one of them as a reservoir of mercury, which it preserves with more safety than any other vessel, being neither subject to overturn, nor to any other accident. We operate with mercury in this apparatus exactly as with water in the one before described; but the bell-glasses must be of smaller diameter, and much stronger; or we may use glass tubes, having their mouths widened, as in Fig. 7.; these are called *eudiometers* by the glass-men who sell them. One of the bell-glasses is represented Fig. 5. A. standing in its place, and what is called a jar is engraved Fig. 6.

The mercurial pneumato-chemical apparatus is necessary in all experiments wherein the disengaged gasses are capable of being absorbed by water, as is frequently the case, especially in all combinations, excepting those of metals, in fermentation, [Pg 308]&c.

SECT. II.
Of the Gazometer.

I give the name of *gazometer* to an instrument which I invented, and caused construct, for the purpose of a kind of bellows, which might furnish an uniform and continued stream of oxygen gas in experiments of fusion. Mr Meusnier and I have since made very considerable corrections and additions, having converted it into what may be called an *universal instrument*, without which it is hardly possible to perform most of the very exact experiments. The name we have given the instrument indicates its intention for measuring the volume or quantity of gas submitted to it for examination.

It consists of a strong iron beam, DE, Pl. VIII. Fig. 1. three feet long, having at each end, D and E, a segment of a circle, likewise strongly constructed of iron, and very firmly joined. Instead of being poised as in ordinary balances, this beam rests, by means of a cylindrical axis of polished steel, F, Fig. 9. upon two large moveable brass friction-wheels, by which the resistance to its motion from friction is considerably diminished, being converted into friction[Pg 309] of the second order. As an additional precaution, the parts of these wheels which support the axis of the beam are covered with plates of polished rock-cristal. The whole of this machinery is fixed to the top of the solid column of wood BC, Fig. 1. To one extremity D of the beam, a scale P for holding weights is suspended by a flat chain, which applies to the curvature of the arc*n*D*o*, in a groove made for the purpose. To the other extremity E of the beam is applied another flat chain, *i k m*, so constructed, as to be incapable of lengthening or shortening, by being less or more charged with weight; to this chain, an iron trivet, with three branches, *a i, c i*, and *h i*, is strongly fixed at *i*, and these branches support a large inverted jar A, of hammered copper, of about 18 inches diameter, and 20 inches deep. The whole of this machine is represented in perspective, Pl. VIII. Fig. 1. and Pl. IX. Fig. 2. and 4. give perpendicular sections, which show its interior structure.

Round the bottom of the jar, on its outside, is fixed (Pl. IX. Fig. 2.) a border divided into compartments 1, 2, 3, 4, &c. intended to receive leaden weights separately represented 1, 2, 3, Fig. 3. These are intended for increasing the weight of the jar when a considerable pressure is requisite, as will be afterwards explained, though such necessity seldom occurs.[Pg 310] The cylindrical jar A is entirely open below, *de*, Pl. IX. Fig. 4.; but is closed above with a copper lid, *a b c*, open at *b f*, and capable of being shut by the cock g. This lid, as may be seen by inspecting the figures, is placed a few inches within the top of the jar to prevent the jar from being ever entirely immersed in the water, and covered over. Were I to have this instrument made over again, I should cause the lid to be considerably more flattened, so as to be almost level. This jar or reservoir of air is contained in the cylindrical copper vessel, LMNO, Pl. VIII. Fig. 1. filled with water.

In the middle of the cylindrical vessel LMNO, Pl. IX. Fig. 4. are placed two tubes *st, xy*, which are made to approach each other at their upper extremities *t y*; these are made of such a length as to rise a little above the upper edge LM of the vessel LMNO, and when the jar *abcde* touches the bottom NO, their upper ends enter about half an inch into the conical hollow *h*, leading to the stop-cock *g*.

The bottom of the vessel LMNO is represented Pl. IX. Fig. 3. in the middle of which a small hollow semispherical cap is soldered, which may be considered as the broad end of a funnel reversed; the two tubes *st, xy*, Fig. 4. are adapted to this cap at *s* and *x*, and by this means communicate with the tubes *mm, nn, oo, pp*, Fig. 3. which are fixed horizontally upon the[Pg 311] bottom of the vessel, and all of which terminate in, and are united by, the spherical cap *sx*. Three of these tubes are continued out of the vessel, as in Pl. VIII. Fig. 1. The first marked in that figure 1, 2, 3, is inserted at its extremity 3, by means of an

134

intermediate stop-cock 4, to the jar V. which stands upon the shelf of a small pneumato-chemical apparatus GHIK, the inside of which is shown Pl. IX. Fig. 1. The second tube is applied against the outside of the vessel LMNO from 6 to 7, is continued at 8, 9, 10, and at 11 is engaged below the jar V. The former of these tubes is intended for conveying gas into the machine, and the latter for conducting small quantities for trials under jars. The gas is made either to flow into or out of the machine, according to the degree of pressure it receives; and this pressure is varied at pleasure, by loading the scale P less or more, by means of weights. When gas is to be introduced into the machine, the pressure is taken off, or even rendered negative; but, when gas is to be expelled, a pressure is made with such degree of force as is found necessary.

The third tube 12, 13, 14, 15, is intended for conveying air or gas to any necessary place or apparatus for combustions, combinations, or any other experiment in which it is required.

To explain the use of the fourth tube, I must enter into some discussions. Suppose the vessel[Pg 312] LMNO, Pl. VIII. Fig. 1. full of water, and the jar A partly filled with gas, and partly with water; it is evident that the weights in the bason P may be so adjusted, as to occasion an exact equilibrium between the weight of the bason and of the jar, so that the external air shall not tend to enter into the jar, nor the gas to escape from it; and in this case the water will stand exactly at the same level both within and without the jar. On the contrary, if the weight in the bason P be diminished, the jar will then press downwards from its own gravity, and the water will stand lower within the jar than it does without; in this case, the included air or gas will suffer a degree of compression above that experienced by the external air, exactly proportioned to the weight of a column of water, equal to the difference of the external and internal surfaces of the water. From these reflections, Mr Meusnier contrived a method of determining the exact degree of pressure to which the gas contained in the jar is at any time exposed. For this purpose, he employs a double glass syphon 19, 20, 21, 22, 23, firmly cemented at 19 and 23. The extremity 19 of this syphon communicates freely with the water in the external vessel of the machine, and the extremity 23 communicates with the fourth tube at the bottom of the cylindrical vessel, and consequently, by means of the perpendicular[Pg 313] tube *st*, Pl. IX. Fig. 4. with the air contained in the jar. He likewise cements, at 16, Pl. VIII. Fig. 1. another glass tube 16, 17, 18, which communicates at 16 with the water in the exterior vessel LMNO, and, at its upper end 18, is open to the external air.

By these several contrivances, it is evident that the water must stand in the tube 16, 17, 18, at the same level with that in the cistern LMNO; and, on the contrary, that, in the branch 19, 20, 21, it must stand higher or lower, according as the air in the jar is subjected to a greater or lesser pressure than the external air. To ascertain these differences, a brass scale divided into inches and lines is fixed between these two tubes. It is readily conceived that, as air, and all other elastic fluids, must increase in weight by compression, it is necessary to know their degree of condensation to be enabled to calculate their quantities, and to convert the measure of their volumes into correspondent weights; and this object is intended to be fulfilled by the contrivance now described.

But, to determine the specific gravity of air or of gasses, and to ascertain their weight in a known volume, it is necessary to know their temperature, as well as the degree of pressure under which they subsist; and this is accomplished by means of a small thermometer, strongly cemented into a brass collet, which screws[Pg 314] into the lid of the jar A. This thermometer is represented separately, Pl. VIII. Fig. 10. and in its place 24, 25, Fig. 1. and Pl. IX. Fig. 4. The bulb is in the inside of the jar A, and its graduated stalk rises on the outside of the lid.

The practice of gazometry would still have laboured under great difficulties, without farther precautions than those above described. When the jar A sinks in the water of the cistern LMNO, it must lose a weight equal to that of the water which it displaces; and consequently the compression which it makes upon the contained air or gas must be proportionally diminished. Hence the gas furnished, during experiments from the machine, will not have the same density towards the end that it had at the beginning, as its specific gravity is continually diminishing. This difference may, it is true, be determined by calculation; but this would have occasioned such mathematical investigations as must have rendered the use of this apparatus both troublesome and difficult. Mr Meusnier has remedied this inconvenience by the following contrivance. A square rod of iron, 26, 27, Pl. VIII. Fig. 1. is raised perpendicular to the middle of the beam DE. This rod passes through a hollow box of brass 28, which opens, and may be filled with lead; and this box is made to slide alongst the rod, by means of a toothed pinion playing in a rack, so as to raise[Pg 315] or lower the box, and to fix it at such places as is judged proper.

When the lever or beam DE stands horizontal, this box gravitates to neither side; but, when the jar A sinks into the cistern LMNO, so as to make the beam incline to that side, it is evident the loaded box 28, which then passes beyond the center of suspension, must gravitate to the side of the jar, and augment its pressure upon the included air. This is increased in proportion as the box is raised towards 27, because the same weight exerts a greater power in proportion to the length of the lever by which it acts. Hence, by moving the box 28 alongst the rod 26, 27, we can augment or diminish the correction it is intended to make upon the pressure of the jar; and both experience and calculation show that this may be made to compensate very exactly for the loss of weight in the jar at all degrees of pressure.

I have not hitherto explained the most important part of the use of this machine, which is the manner of employing it for ascertaining the quantities of the air or gas furnished during experiments. To determine this with the most rigorous precision, and likewise the quantity supplied to the machine from experiments, we fixed to the arc which terminates the arm of the beam E, Pl. VIII. Fig. 1. the brass sector $l\ m$, divided into degrees and half degrees,[Pg 316] which consequently moves in common with the beam; and the lowering of this end of the beam is measured by the fixed index 29, 30, which has a Nonius giving hundredth parts of a degree at its extremity 30.

The whole particulars of the different parts of the above described machine are represented in Plate VIII. as follow.

Fig. 2. Is the flat chain invented by Mr Vaucanson, and employed for suspending the scale or bason P, Fig. 1; but, as this lengthens or shortens according as it is more or less loaded, it would not have answered for suspending the jar A, Fig. 1.

Fig. 5. Is the chain $i\ k\ m$, which in Fig. 1. sustains the jar A. This is entirely formed of plates of polished iron interlaced into each other, and held together by iron pins. This chain does not lengthen in any sensible degree, by any weight it is capable of supporting.

Fig. 6. The trivet, or three branched stirrup, by which the jar A is hung to the balance, with the screw by which it is fixed in an accurately vertical position.

Fig. 3. The iron rod 26, 27, which is fixed perpendicular to the center of the beam, with its box 28.

Fig. 7. & 8. The friction-wheels, with the plates of rock-cristal Z, as points of contact[Pg 317]by which the friction of the axis of the lever of the balance is avoided.

Fig. 4. The piece of metal which supports the axis of the friction-wheels.

Fig. 9. The middle of the lever or beam, with the axis upon which it moves.

Fig. 10. The thermometer for determining the temperature of the air or gas contained in the jar.

When this gazometer is to be used, the cistern or external vessel, LMNO, Pl. VIII. Fig. 1. is to be filled with water to a determinate height, which should be the same in all experiments. The level of the water should be taken when the beam of the balance stands horizontal; this level, when the jar is at the bottom of the cistern, is increased by all the water which it displaces, and is diminished in proportion as the jar rises to its highest elevation. We next endeavour, by repeated trials, to discover at what elevation the box 28 must be fixed, to render the pressure equal in all situations of the beam. I should have said nearly, because this correction is not absolutely rigorous; and differences of a quarter, or even of half a line, are not of any consequence. This height of the box 28 is not the same for every degree of pressure, but varies according as this is of one, two, three, or more inches. All these should be registered with great order and precision.[Pg 318]

We next take a bottle which holds eight or ten pints, the capacity of which is very accurately determined by weighing the water it is capable of containing. This bottle is turned bottom upwards, full of water, in the cistern of the pneumato chemical apparatus GHIK, Fig. 1. and is set on its mouth upon the shelf of the apparatus, instead of the glass jar V, having the extremity 11 of the tube 7, 8, 9, 10, 11, inserted into its mouth. The machine is fixed at zero of pressure, and the degree marked by the index 30 upon the sector *m* /is accurately observed; then, by opening the stop-cock 8, and pressing a little upon the jar A, as much air is forced into the bottle as fills it entirely. The degree marked by the index upon the sector is now observed, and we calculate what number of cubical inches correspond to each degree. We then fill a second and third bottle, and so on, in the same manner, with the same precautions, and even repeat the operation several times with bottles of different sizes, till at last, by accurate attention, we ascertain the exact gage or capacity of the jar A, in all its parts; but it is better to have it formed at first accurately cylindrical, by which we avoid these calculations and estimates.

The instrument I have been describing was constructed with great accuracy and uncommon skill by Mr Meignie junior, engineer and physical[Pg 319] instrument-maker. It is a most valuable instrument, from the great number of purposes to which it is applicable; and, indeed, there are many experiments which are almost impossible to be performed without it. It becomes expensive, because, in many experiments, such as the formation of water and of nitric acid, it is absolutely necessary to employ two of the same machines. In the present advanced state of chemistry, very expensive and complicated instruments are become indispensibly necessary for ascertaining the analysis and synthesis of bodies with the requisite precision as to quantity and proportion; it is certainly proper to endeavour to simplify these, and to render them less costly; but this ought by no means to be attempted at the expence of their conveniency of application, and much less of their accuracy.

SECT. III.
Some other methods of measuring the volume of Gasses.

The gazometer described in the foregoing section is too costly and too complicated for being generally used in laboratories for measuring the gasses, and is not even applicable to every[Pg 320] circumstance of this kind. In numerous series of experiments, more simple and more readily applicable methods must be employed. For this purpose I shall describe the means I used before I was in possession of a gazometer, and which I still use in preference to it in the ordinary course of my experiments.

Suppose that, after an experiment, there is a residuum of gas, neither absorbable by alkali nor water, contained in the upper part of the jar AEF, Pl. IV. Fig. 3. standing on the shelf of a pneumato-chemical apparatus, of which we wish to ascertain the quantity, we must first mark the height to which the mercury or water rises in the jar with great exactness, by means of slips of paper pasted in several parts round the jar. If we have been operating in mercury, we begin by displacing the mercury from the jar, by introducing water in its stead. This is readily done by filling a bottle quite full of water; having stopped it with your finger, turn it up, and introduce its mouth below the edge of the jar; then, turning down its body again, the mercury, by its gravity, falls into the bottle, and the water rises in the jar, and takes the place occupied by the mercury. When this is accomplished, pour so much water into the cistern ABCD as will stand about an inch over the surface of the mercury; then pass the dish BC, Pl. V. Fig. 9. under the jar, and carry it to the[Pg 321] water cistern, Fig. 1. and 2. We here exchange the gas into another jar, which has been previously graduated in the manner to be afterwards described; and we thus judge of the quantity or volume of the gas by means of the degrees which it occupies in the graduated jar.

There is another method of determining the volume of gas, which may either be substituted in place of the one above described, or may be usefully employed as a correction or proof of that method. After the air or gas is exchanged from the first jar, marked with slips of paper, into the graduated jar, turn up the mouth of the marked jar, and fill it with water exactly to the marks EF, Pl. IV. Fig. 3. and by weighing the water we determine the volume of the air or gas it contained, allowing one cubical foot, or 1728 cubical inches, of water for each 70 pounds, French weight.

The manner of graduating jars for this purpose is very easy, and we ought to be provided with several of different sizes, and even several of each size, in case of accidents. Take a tall, narrow, and strong glass jar, and, having filled it with water in the cistern, Pl. V.

Fig. 1. place it upon the shelf ABCD; we ought always to use the same place for this operation, that the level of the shelf may be always exactly similar, by which almost the only error to which this process is liable will be avoided. Then take a narrow[Pg 322] mouthed phial which holds exactly 6 *oz.* 3 *gros* 61 *grs.* of water, which corresponds to 10 cubical inches. If you have not one exactly of this dimension, choose one a little larger, and diminish its capacity to the size requisite, by dropping in a little melted wax and rosin. This bottle serves the purpose of a standard for gaging the jars. Make the air contained in this bottle pass into the jar, and mark exactly the place to which the water has descended; add another measure of air, and again mark the place of the water, and so on, till all the water be displaced. It is of great consequence that, during the course of this operation, the bottle and jar be kept at the same temperature with the water in the cistern; and, for this reason, we must avoid keeping the hands upon either as much as possible; or, if we suspect they have been heated, we must cool them by means of the water in the cistern. The height of the barometer and thermometer during this experiment is of no consequence.

When the marks have been thus ascertained upon the jar for every ten cubical inches, we engrave a scale upon one of its sides, by means of a diamond pencil. Glass tubes are graduated in the same manner for using in the mercurial apparatus, only they must be divided into cubical inches, and tenths of a cubical inch. The bottle used for gaging these must hold[Pg 323] 8 *oz.* 6 *gros* 25 *grs.* of mercury, which exactly corresponds to a cubical inch of that metal.

The method of determining the volume of air or gas, by means of a graduated jar, has the advantage of not requiring any correction for the difference of height between the surface of the water within the jar, and in the cistern; but it requires corrections with respect to the height of the barometer and thermometer. But, when we ascertain the volume of air by weighing the water which the jar is capable of containing, up to the marks EF, it is necessary to make a farther correction, for the difference between the surface of the water in the cistern, and the height to which it rises within the jar. This will be explained in the fifth section of this chapter.

SECT. IV.
Of the method of Separating the different Gasses from each other.

As experiments often produce two, three, or more species of gas, it is necessary to be able to separate these from each other, that we may ascertain the quantity and species of each. Suppose that under the jar A, Pl. IV. Fig. 3. is[Pg 324] contained a quantity of different gasses mixed together, and standing over mercury, we begin by marking with slips of paper, as before directed, the height at which the mercury stands within the glass; then introduce about a cubical inch of water into the jar, which will swim over the surface of the mercury: If the mixture of gas contains any muriatic or sulphurous acid gas, a rapid and considerable absorption will instantly take place, from the strong tendency these two gasses have, especially the former, to combine with, or be absorbed by water. If the water only produces a slight absorption of gas hardly equal to its own bulk, we conclude, that the mixture neither contains muriatic acid, sulphuric acid, or ammoniacal gas, but that it contains carbonic acid gas, of which water only absorbs about its own bulk. To ascertain this conjecture, introduce some solution of caustic alkali, and the carbonic acid gas will be gradually absorbed in the course of a few hours; it combines with the caustic alkali or potash, and the remaining gas is left almost perfectly free from any sensible residuum of carbonic acid gas.

After each experiment of this kind, we must carefully mark the height at which the mercury stands within the jar, by slips of paper pasted on, and varnished over when dry, that they may not be washed off when placed in the water apparatus.[Pg 325] It is likewise necessary to register the difference between the surface of the mercury in the cistern and that in the jar, and the height of the barometer and thermometer, at the end of each experiment.

When all the gas or gasses absorbable by water and potash are absorbed, water is admitted into the jar to displace the mercury; and, as is described in the preceding section, the mercury in the cistern is to be covered by one or two inches of water. After this, the jar is to be transported by means of the flat dish BC, Pl. V. Fig. 9. into the water apparatus; and the quantity of gas remaining is to be ascertained by changing it into a graduated jar. After this, small trials of it are to be made by experiments in little jars, to ascertain nearly the nature of the gas in question. For instance, into a small jar full of the gas, Fig. 8. Pl. V. a lighted taper is introduced; if the taper is not immediately extinguished, we conclude the gas to contain oxygen gas; and, in proportion to the brightness of the flame, we may judge if it contain less or more oxygen gas than atmospheric air contains. If, on the contrary, the taper be instantly extinguished, we have strong reason to presume that the residuum is chiefly composed of azotic gas. If, upon the approach of the taper, the gas takes fire and burns quietly at the surface with a white flame, we conclude it to be[Pg 326] pure hydrogen gas; if this flame is blue, we judge it consists of carbonated hydrogen gas; and, if it takes fire with a sudden deflagration, that it is a mixture of oxygen and hydrogen gas. If, again, upon mixing a portion of the residuum with oxygen gas, red fumes are produced, we conclude that it contains nitrous gas.

These preliminary trials give some general knowledge of the properties of the gas, and nature of the mixture, but are not sufficient to determine the proportions and quantities of the several gasses of which it is composed. For this purpose all the methods of analysis must be employed; and, to direct these properly, it is of great use to have a previous approximation by the above methods. Suppose, for instance, we know that the residuum consists of oxygen and azotic gas mixed together, put a determinate quantity, 100 parts, into a graduated tube of ten or twelve lines diameter, introduce a solution of sulphuret of potash in contact with the gas, and leave them together for some days; the sulphuret absorbs the whole oxygen gas, and leaves the azotic gas pure.

If it is known to contain hydrogen gas, a determinate quantity is introduced into Volta's eudiometer alongst with a known proportion of hydrogen gas; these are deflagrated together by means of the electrical spark; fresh portions of oxygen gas are successively added, till no farther[Pg 327] deflagration takes place, and till the greatest possible diminution is produced. By this process water is formed, which is immediately absorbed by the water of the apparatus; but, if the hydrogen gas contain charcoal, carbonic acid is formed at the same time, which is not absorbed so quickly; the quantity of this is readily ascertained by assisting its absorption, by means of agitation. If the residuum contains nitrous gas, by adding oxygen gas, with which it combines into nitric acid, we can very nearly ascertain its quantity, from the diminution produced by this mixture.

I confine myself to these general examples, which are sufficient to give an idea of this kind of operations; a whole volume would not serve to explain every possible case. It is necessary to become familiar with the analysis of gasses by long experience; we must even acknowledge that they mostly possess such powerful affinities to each other, that we are not always certain of having separated them completely. In these cases, we must vary our experiments in every possible point of view, add new agents to the combination, and keep out others, and continue our trials, till we are certain of the truth and exactitude of our conclusions.[Pg 328]

SECT. V.
Of the necessary corrections upon the volume of the Gasses, according to the pressure of the Atmosphere.

All elastic fluids are compressible or condensible in proportion to the weight with which they are loaded. Perhaps this law, which is ascertained by general experience, may suffer some irregularity when these fluids are under a degree of condensation almost sufficient to reduce them to the liquid state, or when either in a state of extreme rarefaction or condensation; but we seldom approach either of these limits with most of the gasses which we submit to our experiments. I understand this proposition of gasses being compressible, in proportion to their superincumbent weights, as follows:

A barometer, which is an instrument generally known, is, properly speaking, a species of syphon, ABCD, Pl. XII. Fig. 16. whose leg AB is filled with mercury, whilst the leg CD is full of air. If we suppose the branch CD indefinitely continued till it equals the height of our atmosphere, we can readily conceive that the barometer is, in reality, a sort of balance, in which[Pg 329] a column of mercury stands in equilibrium with a column of air of the same weight. But it is unnecessary to prolongate the branch CD to such a height, as it is evident that the barometer being immersed in air, the column of mercury AB will be equally in equilibrium with a column of air of the same diameter, though the leg CD be cut off at C, and the part CD be taken away altogether.

The medium height of mercury in equilibrium with the weight of a column of air, from the highest part of the atmosphere to the surface of the earth is about twenty-eight French inches in the lower parts of the city of Paris; or, in other words, the air at the surface of the earth at Paris is usually pressed upon by a weight equal to that of a column of mercury twenty-eight inches in height. I must be understood in this way in the several parts of this publication when talking of the different gasses, as, for instance, when the cubical foot of oxygen gas is said to weigh 1 *oz.* 4 *gros*, under 28 inches pressure. The height of this column of mercury, supported by the pressure of the air, diminishes in proportion as we are elevated above the surface of the earth, or rather above the level of the sea, because the mercury can only form an equilibrium with the column of air which is above it, and is not in the smallest[Pg 330] degree affected by the air which is below its level.

In what ratio does the mercury in the barometer descend in proportion to its elevation? or, what is the same thing, according to what law or ratio do the several strata of the atmosphere decrease in density? This question, which has exercised the ingenuity of natural philosophers during last century, is considerably elucidated by the following experiment.

If we take the glass syphon ABCDE, Pl. XII. Fig. 17. shut at E, and open at A, and introduce a few drops of mercury, so as to intercept the communication of air between the leg AB and the leg BE, it is evident that the air contained in BCDE is pressed upon, in common with the whole surrounding air, by a weight or column of air equal to 28 inches of mercury. But, if we pour 28 inches of mercury into the leg AB, it is plain the air in the branch BCDE will now be pressed upon by a weight equal to twice 28 inches of mercury, or twice the weight of the atmosphere; and experience shows, that, in this case, the included air, instead of filling the tube from B to E, only occupies from C to E, or exactly one half of the space it filled before. If to this first column of mercury we add two other portions of 28 inches each, in the branch AB, the air in the branch BCDE will be pressed upon by four times the weight of the[Pg 331]atmosphere, or four times the weight of 28 inches of mercury, and it will then only fill the space from D to E, or exactly one quarter of the space it occupied at the commencement of the experiment. From these experiments, which may be infinitely varied, has been deduced as a general law of nature, which seems applicable to all permanently elastic fluids, that they diminish in volume in proportion to the weights with which they are pressed upon; or, in other words, "*the volume of all elastic fluids is in the inverse ratio of the weight by which they are compressed.*"

The experiments which have been made for measuring the heights of mountains by means of the barometer, confirm the truth of these deductions; and, even supposing them in some degree inaccurate, these differences are so extremely small, that they may be reckoned as nullities in chemical experiments. When this law of the compression of elastic fluids is once well understood, it becomes easily applicable to the corrections necessary in pneumato chemical experiments upon the volume of gas, in relation to its pressure. These corrections are of two kinds, the one relative to the variations of the barometer, and the other for the column of water or mercury contained in the jars. I shall endeavour to explain these by examples, beginning with the most simple case.[Pg 332]

Suppose that 100 cubical inches of oxygen gas are obtained at 10° (54.5°) of the thermometer, and at 28 inches 6 lines of the barometer, it is required to know what volume the 100 cubical inches of gas would occupy, under the pressure of 28 inches[58], and what is the exact weight of the 100 inches of oxygen gas? Let the unknown volume, or the number of inches this gas would occupy at 28 inches of the barometer, be expressed by x; and, since the volumes are in the inverse ratio of their superincumbent weights, we have the following statement: 100 cubical inches is to x inversely as 28.5 inches of pressure is to 28.0 inches; or directly 28 : 28.5 :: 100 : x = 101.786—cubical inches, at 28 inches barometrical pressure; that is to say, the same gas or air which at 28.5 inches of the barometer occupies 100 cubical inches of volume, will occupy 101.786 cubical inches when the barometer is at 28 inches. It is equally easy to calculate the weight of this gas, occupying 100 cubical inches, under 28.5 inches of barometrical pressure; for, as it corresponds[Pg 333] to 101.786 cubical inches at

140

the pressure of 28, and as, at this pressure, and at 10° (54.5°) of temperature, each cubical inch of oxygen gas weighs half a grain, it follows, that 100 cubical inches, under 28.5 barometrical pressure, must weigh 50.893 grains. This conclusion might have been formed more directly, as, since the volume of elastic fluids is in the inverse ratio of their compression, their weights must be in the direct ratio of the same compression: Hence, since 100 cubical inches weigh 50 grains, under the pressure of 28 inches, we have the following statement to determine the weight of 100 cubical inches of the same gas as 28.5 barometrical pressure, 28 : 50 :: 28.5 : x, the unknown quantity, = 50.893.

The following case is more complicated: Suppose the jar A, Pl. XII. Fig. 18. to contain a quantity of gas in its upper part ACD, the rest of the jar below CD being full of mercury, and the whole standing in the mercurial bason or reservoir GHIK, filled with mercury up to EF, and that the difference between the surface CD of the mercury in the jar, and EF, that in the cistern, is six inches, while the barometer stands at 27.5 inches. It is evident from these data, that the air contained in ACD is pressed upon by the weight of the atmosphere, diminished by the weight of the column of mercury CE, or by 27.5 - 6 = 21.5 inches of barometrical[Pg 334] pressure. This air is therefore less compressed than the atmosphere at the mean height of the barometer, and consequently occupies more space than it would occupy at the mean pressure, the difference being exactly proportional to the difference between the compressing weights. If, then, upon measuring the space ACD, it is found to be 120 cubical inches, it must be reduced to the volume which it would occupy under the mean pressure of 28 inches. This is done by the following statement: 120 : x, the unknown volume, :: 21.5 : 28 inversely; this gives x = 120 × 21.5 / 28 = 92.143 cubical inches.

In these calculations we may either reduce the height of the mercury in the barometer, and the difference of level in the jar and bason, into lines or decimal fractions of the inch; but I prefer the latter, as it is more readily calculated. As, in these operations, which frequently recur, it is of great use to have means of abbreviation, I have given a table in the appendix for reducing lines and fractions of lines into decimal fractions of the inch.

In experiments performed in the water-apparatus, we must make similar corrections to procure rigorously exact results, by taking into account, and making allowances for the difference of height of the water within the jar above the surface of the water in the cistern. But, as the[Pg 335] pressure of the atmosphere is expressed in inches and lines of the mercurial barometer, and, as homogeneous quantities only can be calculated together, we must reduce the observed inches and lines of water into correspondent heights of the mercury. I have given a table in the appendix for this conversion, upon the supposition that mercury is 13.5681 times heavier than water.

SECT. VI.
Of Corrections relative to the Degrees of the Thermometer.

In ascertaining the weight of gasses, besides reducing them to a mean of barometrical pressure, as directed in the preceding section, we must likewise reduce them to a standard thermometrical temperature; because, all elastic fluids being expanded by heat, and condensed by cold, their weight in any determinate volume is thereby liable to considerable alterations. As the temperature of 10° (54.5°) is a medium between the heat of summer and the cold of winter, being the temperature of subterraneous places, and that which is most easily approached to at all seasons, I have chosen that degree as a mean to which I reduce air or gas in this species of calculation.[Pg 336]

Mr de Luc found that atmospheric air was increased 1/215 part of its bulk, by each degree of a mercurial thermometer, divided into 81 degrees, between the freezing and boiling points; this gives 1/211 part for each degree of Reaumur's thermometer, which is divided into 80 degrees between these two points. The experiments of Mr Monge seem to make this dilatation less for hydrogen gas, which he thinks is only dilated 1/180. We have not any exact experiments hitherto published respecting the ratio of dilatation of the other gasses; but, from the trials which have been made, their dilatation seems to differ little from that of atmospheric air. Hence I may take for granted, till farther experiments give us better information upon this subject, that atmospherical air is dilated 1/210 part, and hydrogen gas 1/190 part for each degree of the thermometer; but, as there is still great uncertainty upon this point, we ought always to operate in a temperature as near as possible to the standard of 10°, (54.5°) by this means any errors in correcting the weight or volume of gasses by reducing them to the common standard, will become of little moment.

The calculation for this correction is extremely easy. Divide the observed volume of air by 210, and multiply the quotient by the degrees of temperature above or below 10°[Pg 337] (54.5°). This correction is negative when the actual temperature is above the standard, and positive when below. By the use of logarithmical tables this calculation is much facilitated[59].

SECT. VII.
Example for calculating the Corrections relative to the Variations of Pressure and Temperature.
CASE.

In the jar A, Pl. IV. Fig. 3. standing in a water apparatus, is contained 353 cubical inches of air; the surface of the water within the jar at EF is 4-1/2 inches above the water in the cistern, the barometer is at 27 inches 9-1/2 lines, and the thermometer at 15° (65.75°). Having burnt a quantity of phosphorus in the air, by which concrete phosphoric acid is produced, the air after the combustion occupies 295 cubical[Pg 338] inches, the water within the jar stands 7 inches above that in the cistern, the barometer is at 27 inches 9-1/4 lines, and the thermometer at 16° (68°). It is required from these data to determine the actual volume of air before and after combustion, and the quantity absorbed during the process.

Calculation before Combustion.

The air in the jar before combustion was 353 cubical inches, but it was only under a barometrical pressure of 27 inches 9-1/2 lines; which, reduced to decimal fractions by Tab. I. of the Appendix, gives 27.79167 inches; and from this we must deduct the difference of 4-1/2 inches of water, which, by Tab. II. corresponds to 0.33166 inches of the barometer; hence the real pressure of the air in the jar is 27.46001. As the volume of elastic fluids diminish in the inverse ratio of the compressing weights, we have the following statement to reduce the 353 inches to the volume the air would occupy at 28 inches barometrical pressure.

353 : x, the unknown volume, :: 27.46001 : 28. Hence, $x = 353 \times 27.46001 / 28 = 346.192$ cubical inches, which is the volume the same quantity of air would have occupied at 28 inches of the barometer.[Pg 339]

The 210th part of this corrected volume is 1.65, which, for the five degrees of temperature above the standard gives 8.255 cubical inches; and, as this correction is subtractive, the real corrected volume of the air before combustion is 337.942 inches.

Calculation after Combustion.

By a similar calculation upon the volume of air after combustion, we find its barometrical pressure 27.77083 - 0.51593 = 27.25490. Hence, to have the volume of air under the pressure of 28 inches, 295 : x :: 27.77083 : 28 inversely; or, $x = 295 \times 27.25490 / 28 = 287.150$. The 210th part of this corrected volume is 1.368, which, multiplied by 6 degrees of thermometrical difference, gives the subtractive correction for temperature 8.208, leaving the actual corrected volume of air after combustion 278.942 inches.

Result.

The corrected volume before combustion	337.942
Ditto remaining after combustion	278.942
	—
Volume absorbed during combustion	59.000

[Pg 340]

SECT. VIII.
Method of determining the Absolute Gravity of the different Gasses.

Take a large balloon A, Pl. V. Fig. 10. capable of holding 17 or 18 pints, or about half a cubical foot, having the brass cap *bcde* strongly cemented to its neck, and to which the tube and stop-cock *f g* is fixed by a tight screw. This apparatus is connected by the double screw represented separately at Fig. 12. to the jar BCD, Fig. 10. which must be some pints larger in dimensions than the balloon. This jar is open at top, and is furnished with the brass cap *h i*, and stop-cock *l m*. One of these slop-cocks is represented separately at Fig. 11.

We first determine the exact capacity of the balloon by filling it with water, and weighing it both full and empty. When emptied of water, it is dried with a cloth introduced through its neck *d e*, and the last remains of moisture are removed by exhausting it once or twice in an air-pump.

When the weight of any gas is to be ascertained, this apparatus is used as follows: Fix the balloon A to the plate of an air-pump by means of the screw of the stop-cock *f g*, which is[Pg 341] left open; the balloon is to be exhausted as completely as possible, observing carefully the degree of exhaustion by means of the barometer attached to the air-pump. When the vacuum is formed, the stop-cock *f g* is shut, and the weight of the balloon determined with the most scrupulous exactitude. It is then fixed to the jar BCD, which we suppose placed in water in the shelf of the pneumato chemical apparatus Fig. 1.; the jar is to be filled with the gas we mean to weigh, and then, by opening the stop-cocks *f g* and *l m*, the gas ascends into the balloon, whilst the water of the cistern rises at the same time into the jar. To avoid very troublesome corrections, it is necessary, during this first part of the operation, to sink the jar in the cistern till the surfaces of the water within the jar and without exactly correspond. The stop-cocks are again shut, and the balloon being unscrewed from its connection with the jar, is to be carefully weighed; the difference between this weight and that of the exhausted balloon is the precise weight of the air or gas contained in the balloon. Multiply this weight by 1728, the number of cubical inches in a cubical foot, and divide the product by the number of cubical inches contained in the balloon, the quotient is the weight of a cubical foot of the gas or air submitted to experiment.[Pg 342]

Exact account must be kept of the barometrical height and temperature of the thermometer during the above experiment; and from these the resulting weight of a cubical foot is easily corrected to the standard of 28 inches and 10°, as directed in the preceding section. The small portion of air remaining in the balloon after forming the vacuum must likewise be attended to, which is easily determined by the barometer attached to the air-pump. If that barometer, for instance, remains at the hundredth part of the height it stood at before the vacuum was formed, we conclude that one hundredth part of the air originally contained remained in the balloon, and consequently that only 99/100 of gas was introduced from the jar into the balloon.

FOOTNOTES:

[58]According to the proportion of 114 to 107, given between the French and English foot, 28 inches of the French barometer are equal to 29.83 inches of the English. Directions will be found in the appendix for converting all the French weights and measures used in this work into corresponding English denominations.—E.

[59]When Fahrenheit's thermometer is employed, the dilatation by each degree must be smaller, in the proportion of 1 to 2.25, because each degree of Reaumur's scale contains 2.25 degrees of Fahrenheit; hence we must divide by 472.5, and finish the rest of the calculation as above.—E.

[Pg 343]

CHAP. III.
Description of the Calorimeter, or Apparatus for measuring Caloric.

The calorimeter, or apparatus for measuring the relative quantities of heat contained in bodies, was described by Mr de la Place and me in the Memoirs of the Academy for 1780, p. 355. and from that Essay the materials of this chapter are extracted.

If, after having cooled any body to the freezing point, it be exposed in an atmosphere of 25° (88.25°), the body will gradually become heated, from the surface inwards, till at last it acquire the same temperature with the surrounding air. But, if a piece of ice be placed in the same situation, the circumstances are quite different; it does not approach in the smallest degree towards the temperature of the circumambient air, but remains constantly at Zero (32°), or the temperature of melting ice, till the last portion of ice be completely melted.

This phenomenon is readily explained; as, to melt ice, or reduce it to water, it requires to be combined with a certain portion of caloric;[Pg 344] the whole caloric attracted from the surrounding bodies, is arrested or fixed at the surface or external layer of ice which it is employed to dissolve, and combines with it to form water; the next quantity of caloric combines with the second layer to dissolve it into water, and so on successively till the whole ice be dissolved or converted into water by combination with caloric, the very last atom still remaining at its former temperature, because the caloric has never penetrated so far as long as any intermediate ice remained to melt.

Upon these principles, if we conceive a hollow sphere of ice at the temperature of Zero (32°) placed in an atmosphere 10° (54.5°), and containing a substance at any degree of temperature above freezing, it follows, 1st, That the heat of the external atmosphere cannot penetrate into the internal hollow of the sphere of ice; 2dly, That the heat of the body placed in the hollow of the sphere cannot penetrate outwards beyond it, but will be stopped at the internal surface, and continually employed to melt successive layers of ice, until the temperature of the body be reduced to Zero (32°), by having all its superabundant caloric above that temperature carried off by the ice. If the whole water, formed within the sphere of ice during the reduction of the temperature of the included body to Zero, be carefully collected, the weight[Pg 345] of the water will be exactly proportional to the quantity of caloric lost by the body in passing from its original temperature to that of melting ice; for it is evident that a double quantity of caloric would have melted twice the quantity of ice; hence the quantity of ice melted is a very exact measure of the quantity of caloric employed to produce that effect, and consequently of the quantity lost by the only substance that could possibly have supplied it.

I have made this supposition of what would take place in a hollow sphere of ice, for the purpose of more readily explaining the method used in this species of experiment, which was first conceived by Mr de la Place. It would be difficult to procure such spheres of ices and inconvenient to make use of them when got; but, by means of the following apparatus, we have remedied that defect. I acknowledge the name of Calorimeter, which I have given it, as derived partly from Greek and partly from Latin, is in some degree open to criticism; but, in matters of science, a slight deviation from strict etymology, for the sake of giving distinctness of idea, is excusable; and I could not derive the name entirely from Greek without approaching too near to the names of known instruments employed for other purposes.

The calorimeter is represented in Pl. VI. It is shown in perspective at Fig. 1. and its interior[Pg 346] structure is engraved in Fig. 2. and 3.; the former being a horizontal, and the latter a perpendicular section. Its capacity or cavity is divided into three parts, which, for better distinction, I shall name the interior, middle, and external cavities. The interior cavity f fff, Fig. 4. into which the substances submitted to experiment are put, is composed of a grating or cage of iron wire, supported by several iron bars; its opening or mouth LM, is covered by the lid HG, of the same materials. The middle cavity $b\ b\ b\ b$, Fig. 2. and 3. is intended to contain the ice which surrounds the interior cavity, and which is to be melted by the caloric of the substance employed in the experiment. The ice is supported by the grate $m\ m$ at the bottom of the cavity, under which is placed the sieve $n\ n$. These two are represented separately in Fig. 5. and 6.

In proportion as the ice contained in the middle cavity is melted, by the caloric disengaged from the body placed in the interior cavity, the water runs through the grate and sieve, and falls through the conical funnel $c\ c\ d$, Fig. 3. and tube $x\ y$, into the receiver F, Fig. 1. This water may be retained or let out at pleasure, by means of the stop-cock u. The external cavity $a\ a\ a\ a$, Fig. 2. and 3. is filled with ice, to prevent any effect upon the ice in the middle cavity from the heat of the surrounding air, and[Pg 347] the water produced from it is carried off through the pipe ST, which shuts by means of the stop-cock r. The whole machine is covered by the lid FF, Fig. 7. made of tin painted with oil colour, to prevent rust.

When this machine is to be employed, the middle cavity *b b b b*, Fig. 2. and 3., the lid GH, Fig. 4. of the interior cavity, the external cavity *a a a a*, Fig. 2. and 3. and the general lid FF, Fig. 7. are all filled with pounded ice, well rammed, so that no void spaces remain, and the ice of the middle cavity is allowed to drain. The machine is then opened, and the substance submitted to experiment being placed in the interior cavity, it is instantly closed. After waiting till the included body is completely cooled to the freezing point, and the whole melted ice has drained from the middle cavity, the water collected in the vessel F, Fig. 1. is accurately weighed. The weight of the water produced during the experiment is an exact measure of the caloric disengaged during the cooling of the included body, as this substance is evidently in a similar situation with the one formerly mentioned as included in a hollow sphere of ice; the whole caloric disengaged is stopped by the ice in the middle cavity, and that ice is preserved from being affected by any other heat by means of the ice contained in the general lid, Fig. 7. and in the external cavity. Experiments[Pg 348] of this kind last from fifteen to twenty hours; they are sometimes accelerated by covering up the substance in the interior cavity with well drained ice, which hastens its cooling.

The substances to be operated upon are placed in the thin iron bucket, Fig. 8. the cover of which has an opening fitted with a cork, into which a small thermometer is fixed. When we use acids, or other fluids capable of injuring the metal of the instruments, they are contained in the matras, Fig. 10. which has a similar thermometer in a cork fitted to its mouth, and which stands in the interior cavity upon the small cylindrical support RS, Fig. 10.

It is absolutely requisite that there be no communication between the external and middle cavities of the calorimeter, otherwise the ice melted by the influence of the surrounding air, in the external cavity, would mix with the water produced from the ice of the middle cavity, which would no longer be a measure of the caloric lost by the substance submitted to experiment.

When the temperature of the atmosphere is only a few degrees above the freezing point, its heat can hardly reach the middle cavity, being arrested by the ice of the cover, Fig. 7. and of the external cavity; but, if the temperature of the air be under the degree of freezing, it might cool the ice contained in the middle cavity, by[Pg 349] causing the ice in the external cavity to fall, in the first place, below zero ($32°$). It is therefore essential that this experiment be carried on in a temperature somewhat above freezing: Hence, in time of frost, the calorimeter must be kept in an apartment carefully heated. It is likewise necessary that the ice employed be not under zero ($32°$); for which purpose it must be pounded, and spread out thin for some time, in a place of a higher temperature.

The ice of the interior cavity always retains a certain quantity of water adhering to its surface, which may be supposed to belong to the result of the experiment; but as, at the beginning of each experiment, the ice is already saturated with as much water as it can contain, if any of the water produced by the caloric should remain attached to the ice, it is evident, that very nearly an equal quantity of what adhered to it before the experiment must have run down into the vessel F in its stead; for the inner surface of the ice in the middle cavity is very little changed during the experiment.

By any contrivance that could be devised, we could not prevent the access of the external air into the interior cavity when the atmosphere was $9°$ or $10°$ ($52°$ or $54°$) above zero. The air confined in the cavity being in that case specifically heavier than the external air, escapes downwards through the pipe *x y*, Fig. 3, and is[Pg 350] replaced by the warmer external air, which, giving out its caloric to the ice, becomes heavier, and sinks in its turn; thus a current of air is formed through the machine, which is the more rapid in proportion as the external air exceeds the internal in temperature. This current of warm air must melt a part of the ice, and injure the accuracy of the experiment: We may, in a great degree, guard against this source of error by keeping the stop-cock *u* continually shut; but it is better to operate only when the temperature of the external air does not exceed $3°$, or at most $4°$, ($39°$ to $41°$); for we have observed, that, in this case, the melting of the interior ice by the atmospheric air is perfectly insensible; so that we may answer for the accuracy of our experiments upon the specific heat of bodies to a fortieth part.

We have caused make two of the above described machines; one, which is intended for such experiments as do not require the interior air to be renewed, is precisely formed according to the description here given; the other, which answers for experiments upon combustion, respiration, &c. in which fresh quantities of air are indispensibly necessary, differs from the former in having two small tubes in the two lids, by which a current of atmospheric air may be blown into the interior cavity of the machine.[Pg 351]

It is extremely easy, with this apparatus, to determine the phenomena which occur in operations where caloric is either disengaged or absorbed. If we wish, for instance, to ascertain the quantity of caloric which is disengaged from a solid body in cooling a certain number of degrees, let its temperature be raised to 80° (212°); it is then placed in the interior cavity $ffff$, Fig. 2. and 3. of the calorimeter, and allowed to remain till we are certain that its temperature is reduced to zero (32°); the water produced by melting the ice during its cooling is collected, and carefully weighed; and this weight, divided by the volume of the body submitted to experiment, multiplied into the degrees of temperature which it had above zero at the commencement of the experiment, gives the proportion of what the English philosophers call specific heat.

Fluids are contained in proper vessels, whose specific heat has been previously ascertained, and operated upon in the machine in the same manner as directed for solids, taking care to deduct, from the quantity of water melted during the experiment, the proportion which belongs to the containing vessel.

If the quantity of caloric disengaged during the combination of different substances is to be determined, these substances are to be previously reduced to the freezing degree by keeping[Pg 352] them a sufficient time surrounded with pounded ice; the mixture is then to be made in the inner cavity of the calorimeter, in a proper vessel likewise reduced to zero (32°); and they are kept inclosed till the temperature of the combination has returned to the same degree: The quantity of water produced is a measure of the caloric disengaged during the combination.

To determine the quantity of caloric disengaged during combustion, and during animal respiration, the combustible bodies are burnt, or the animals are made to breathe in the interior cavity, and the water produced is carefully collected. Guinea pigs, which resist the effects of cold extremely well, are well adapted for this experiment. As the continual renewal of air is absolutely necessary in such experiments, we blow fresh air into the interior cavity of the calorimeter by means of a pipe destined for that purpose, and allow it to escape through another pipe of the same kind; and that the heat of this air may not produce errors in the results of the experiments, the tube which conveys it into the machine is made to pass through pounded ice, that it may be reduced to zero (32°) before it arrives at the calorimeter. The air which escapes must likewise be made to pass through a tube surrounded with ice, included in the interior cavity of the machine, and the water which is produced must make a part of what is[Pg 353] collected, because the caloric disengaged from this air is part of the product of the experiment.

It is somewhat more difficult to determine the specific caloric contained in the different gasses, on account of their small degree of density; for, if they are only placed in the calorimeter in vessels like other fluids, the quantity of ice melted is so small, that the result of the experiment becomes at best very uncertain. For this species of experiment we have contrived to make the air pass through two metallic worms, or spiral tubes; one of these, through which the air passes, and becomes heated in its way to the calorimeter, is contained in a vessel full of boiling water, and the other, through which the air circulates within the calorimeter to disengage its caloric, is placed in the interior cavity, $ffff$, of that machine. By means of a small thermometer placed at one end of the second worm, the temperature of the air, as it enters the calorimeter, is determined, and its temperature in getting out of the interior cavity is found by another thermometer placed at the other end of the worm. By this contrivance we are enabled to ascertain the quantity of ice melted by determinate quantities of air or gas, while losing a certain number of degrees of temperature, and, consequently, to determine their several degrees of specific caloric. The[Pg 354] same apparatus, with some particular precautions, may be employed to ascertain the quantity of caloric disengaged by the condensation of the vapours of different liquids.

The various experiments which may be made with the calorimeter do not afford absolute conclusions, but only give us the measure of relative quantities; we have therefore to fix a unit, or standard point, from whence to form a scale of the several results. The quantity of caloric necessary to melt a pound of ice has been chosen as this unit; and, as it requires a pound of water of the temperature of 60° (167°) to melt a pound of ice, the quantity of caloric expressed by our unit or standard point is what raises a pound of water from zero (32°) to 60° (167°). When this unit is once determined, we have only to express the quantities of caloric disengaged from different bodies by cooling a certain number of degrees, in analogous values: The following is an easy mode of calculation for this purpose, applied to one of our earliest experiments.

We took 7 *lib*. 11 *oz*. 2 *gros* 36 *grs*. of plate-iron, cut into narrow slips, and rolled up, or expressing the quantity in decimals, 7.7070319. These, being heated in a bath of boiling water to about 78° (207.5°), were quickly introduced into the interior cavity of the calorimeter: At[Pg 355] the end of eleven hours, when the whole quantity of water melted from the ice had thoroughly drained off, we found that 1.109795 pounds of ice were melted. Hence, the caloric disengaged from the iron by cooling 78° (175.5°) having melted 1.109795 pounds of ice, how much would have been melted by cooling 60° (135°)? This question gives the following statement in direct proportion, 78 : 1.109795 :: 60 : x = 0.85369. Dividing this quantity by the weight of the whole iron employed, viz. 7.7070319, the quotient 0.110770 is the quantity of ice which would have been melted by one pound of iron whilst cooling through 60° (135°) of temperature.

Fluid substances, such as sulphuric and nitric acids, &c. are contained in a matras, Pl. VI. Fig. 9. having a thermometer adapted to the cork, with its bulb immersed in the liquid. The matras is placed in a bath of boiling water, and when, from the thermometer, we judge the liquid is raised to a proper temperature, the matras is placed in the calorimeter. The calculation of the products, to determine the specific caloric of these fluids, is made as above directed, taking care to deduct from the water obtained the quantity which would have been produced by the matras alone, which must be ascertained by a previous experiment. The[Pg 356] table of the results obtained by these experiments is omitted, because not yet sufficiently complete, different circumstances having occasioned the series to be interrupted; it is not, however, lost sight of; and we are less or more employed upon the subject every winter.

[Pg 357]

CHAP. IV.
Of Mechanical Operations for Division of Bodies.
SECT. I.
Of Trituration, Levigation, and Pulverization.

These are, properly speaking, only preliminary mechanical operations for dividing and separating the particles of bodies, and reducing them into very fine powder. These operations can never reduce substances into their primary, or elementary and ultimate particles; they do not even destroy the aggregation of bodies; for every particle, after the most accurate trituration, forms a small whole, resembling the original mass from which it was divided. The real chemical operations, on the contrary, such as solution, destroy the aggregation of bodies, and separate their constituent and integrant particles from each other.[Pg 358]

Brittle substances are reduced to powder by means of pestles and mortars. These are of brass or iron, Pl. I. Fig. 1.; of marble or granite, Fig. 2.; of lignum vitae, Fig. 3.; of glass, Fig. 4.; of agate, Fig. 5.; or of porcellain, Fig. 6. The pestles for each of these are represented in the plate, immediately below the mortars to which they respectively belong, and are made of hammered iron or brass, of wood, glass, porcellain, marble, granite, or agate, according to the nature of the substances they are intended to triturate. In every laboratory, it is requisite to have an assortment of these utensils, of various sizes and kinds: Those of porcellain and glass can only be used for rubbing substances to powder, by a dexterous use of the pestle round the sides of the mortar, as it would be easily broken by reiterated blows of the pestle.

The bottom of mortars ought to be in the form of a hollow sphere, and their sides should have such a degree of inclination as to make the substances they contain fall back to the bottom when the pestle is lifted, but not so perpendicular as to collect them too much together, otherwise too large a quantity would get below the pestle, and prevent its operation. For this reason, likewise, too large a quantity of the substance to be powdered ought not to be put into the mortar at one time; and we must from[Pg 359] time to time get rid of the particles already reduced to powder, by means of sieves to be afterwards described.

The most usual method of levigation is by means of a flat table ABCD, Pl. 1. Fig. 7. of porphyry, or other stone of similar hardness, upon which the substance to be reduced to powder is spread, and is then bruised and rubbed by a muller M, of the same hard materials, the bottom of which is made a small portion of a large sphere; and, as the muller tends continually to drive the substances towards the sides of the table, a thin flexible knife, or spatula of iron, horn, wood, or ivory, is used for bringing them back to the middle of the stone.

In large works, this operation is performed by means of large rollers of hard stone, which turn upon each other, either horizontally, in the way of corn-mills, or by one vertical roller turning upon a flat stone. In the above operations, it is often requisite to moisten the substances a little, to prevent the fine powder from flying off.

There are many bodies which cannot be reduced to powder by any of the foregoing methods; such are fibrous substances, as woods; such as are tough and elastic, as the horns of animals, elastic gum, &c. and the malleable metals which flatten under the pestle, instead of being reduced to powder. For reducing the[Pg 360] woods to powder, rasps, as Pl. I. Fig. 8. are employed; files of a finer kind are used for horn, and still finer, Pl. 1. Fig. 9. and 10. for metals.

Some of the metals, though not brittle enough to powder under the pestle, are too soft to be filed, as they clog the file, and prevent its operation. Zinc is one of these, but it may be powdered when hot in a heated iron mortar, or it may be rendered brittle, by alloying it with a small quantity of mercury. One or other of these methods is used by fire-work makers for producing a blue flame by means of zinc. Metals may be reduced into grains, by pouring them when melted into water, which serves very well when they are not wanted in fine powder.

Fruits, potatoes, &c. of a pulpy and fibrous nature may be reduced to pulp by means of the grater, Pl. 1. Fig. 11.

The choice of the different substances of which these instruments are made is a matter of importance; brass or copper are unfit for operations upon substances to be used as food or in pharmacy; and marble or metallic instruments must not be used for acid substances; hence mortars of very hard wood, and those of porcelain, granite, or glass, are of great utility in many operations.[Pg 361]

SECT. II.
Of Sifting and Washing Powdered Substances.

None of the mechanical operations employed for reducing bodies to powder is capable of producing it of an equal degree of fineness throughout; the powder obtained by the longest and most accurate trituration being still an assemblage of particles of various sizes. The coarser of these are removed, so as only to leave the finer and more homogeneous particles by means of sieves, Pl. I. Fig. 12. 13. 14. 15. of different finenesses, adapted to the particular purposes they are intended for; all the powdered matter which is larger than the intestices of the sieve remains behind, and is again submitted to the pestle, while the finer pass through. The sieve Fig. 12. is made of hair-cloth, or of silk gauze; and the one represented Fig. 13. is of parchment pierced with round holes of a proper size; this latter is employed in the manufacture of gun-powder. When very subtile or valuable materials are to be sifted, which are easily dispersed, or when the finer parts of the powder may be hurtful, a compound sieve, Fig. 15. is made use of, which consists of the sieve ABCD, with a lid EF, and receiver GH; these three[Pg 362] parts are represented as joined together for use, Fig. 14.

There is a method of procuring powders of an uniform fineness, considerably more accurate than the sieve; but it can only be used with such substances as are not acted upon by water. The powdered substance is mixed and agitated with water, or other convenient fluid; the liquor is allowed to settle for a few moments, and is then decanted off; the coarsest powder remains at the bottom of the vessel, and the finer passes over with the liquid. By repeated decantations in this manner, various sediments are obtained of different degrees of fineness; the last sediment, or that which remains longed suspended in the liquor, being the finest. This process may likewise be used with advantage for separating substances of different degrees of specific gravity, though of the same fineness; this last is chiefly employed in mining, for separating the heavier metallic ores from the lighter earthy matters with which they are mixed.

In chemical laboratories, pans and jugs of glass or earthen ware are employed for this operation; sometimes, for decanting the liquor without disturbing the sediment, the glass syphon ABCHI, Pl. II. Fig. 11. is used, which may be supported by means of the perforated board DE, at the proper depth in the vessel FG, to draw off all the liquor required into the[Pg 363] receiver LM. The principles and application of this useful instrument are so well known as to need no explanation.

SECT. III.
Of Filtration.

A filtre is a species of very fine sieve, which is permeable to the particles of fluids, but through which the particles of the finest powdered solids are incapable of passing; hence its use in separating fine powders from suspension in fluids. In pharmacy, very close and fine woollen cloths are chiefly used for this operation; these are commonly formed in a conical shape, Pl. II. Fig. 2. which has the advantage of uniting all the liquor which drains through into a point A, where it may be readily collected in a narrow mouthed vessel. In large pharmaceutical laboratories, this filtring bag is streached upon a wooden stand, Pl. II. Fig. 1.

For the purposes of chemistry, as it is requisite to have the filtres perfectly clean, unsized paper is substituted instead of cloth or flannel; through this substance, no solid body, however finely it be powdered, can penetrate, and fluids percolate through it with the greatest readiness.[Pg 364] As paper breaks easily when wet, various methods of supporting it are used according to circumstances. When a large quantity of fluid is to be filtrated, the paper is supported by the frame of wood, Pl. II. Fig. 3. ABCD, having a piece of coarse cloth stretched over it, by means of iron-hooks. This cloth must be well cleaned each time it is used, or even new cloth must be employed, if there is reason to suspect its being impregnated with any thing which can injure the subsequent operations. In ordinary operations, where moderate quantities of fluid are to be filtrated, different kinds of glass funnels are used for supporting the paper, as represented Pl. II. Fig. 5. 6. and 7. When several filtrations must be carried on at once, the board or shelf AB, Fig. 9. supported upon stands C and D, and pierced with round holes, is very convenient for containing the funnels.

Some liquors are so thick and clammy, as not to be able to penetrate through paper without some previous preparation, such as clarification by means of white of eggs, which being mixed with the liquor, coagulates when brought to boil, and, entangling the greater part of the impurities of the liquor, rises with them to the surface in the state of scum. Spiritous liquors may be clarified in the same manner by means of isinglass dissolved in water, which coagulates[Pg 365] by the action of the alkohol without the assistance of heat.

As most of the acids are produced by distillation, and are consequently clear, we have rarely any occasion to filtrate them; but if, at any time, concentrated acids require this operation, it is impossible to employ paper, which would be corroded and destroyed by the acid. For this purpose, pounded glass, or rather quartz or rock-cristal, broke in pieces and grossly powdered, answers very well; a few of the larger pieces are put in the neck of the funnel; these are covered with the smaller pieces, the finer powder is placed over all, and the acid is poured on at top. For the ordinary purposes of society, river-water is frequently filtrated by means of clean washed sand, to separate its impurities.

SECT. IV.
Of Decantation.

This operation is often substituted instead of filtration for separating solid particles which are diffused through liquors. These are allowed to settle in conical vessels, ABCDE, Pl. II. Fig. 10. the diffused matters gradually subside, and the[Pg 366] clear fluid is gently poured off. If the sediment be extremely light, and apt to mix again with the fluid by the slightest motion, the syphon, Fig. 11. is used, instead of decantation, for drawing off the clear fluid.

In experiments, where the weight of the precipitate must be rigorously ascertained, decantation is preferable to filtration, providing the precipitate be several times washed in a considerable proportion of water. The weight of the precipitate may indeed be ascertained, by carefully weighing the filtre before and after the operation; but, when the quantity of precipitate is small, the different proportions of moisture retained by the paper, in a greater or lesser degree of exsiccation, may prove a material source of error, which ought carefully to be guarded against.

[Pg 367]

CHAP. V.
Of Chemical Means for separating the Particles of Bodies from each other; without Decomposition, and for uniting them again.

I have already shown that there are two methods of dividing the particles of bodies, the *mechanical* and *chemical*. The former only separates a solid mass into a great number of smaller masses; and for these purposes various species of forces are employed, according to circumstances, such as the strength of man or of animals, the weight of water applied through the means of hydraulic engines, the expansive power of steam, the force of the wind, &c. By all these mechanical powers, we can never reduce substances into powder beyond a certain degree of fineness; and the smallest particle produced in this way, though it seems very minute to our organs, is still in fact a mountain, when compared with the ultimate elementary particles of the pulverized substance.

The chemical agents, on the contrary, divide bodies into their primitive particles. If, for instance, a neutral salt be acted upon by these, it is divided, as far as is possible, without ceasing to be a neutral salt. In this Chapter, I mean to[Pg 368] give examples of this kind of division of bodies, to which I shall add some account of the relative operations.

SECT. I.
Of the Solution of Salts.

In chemical language, the terms of *solution* and *dissolution* have long been confounded, and have very improperly been indiscriminately employed for expressing both the division of the particles of a salt in a fluid, such as water, and the division of a metal in an acid. A few reflections upon the effects of these two operations will suffice to show that they ought not to be confounded together. In the solution of salts, the saline particles are only separated from each other, whilst neither the salt nor the water are at all decomposed; we are able to recover both the one and the other in the same quantity as before the operation. The same thing takes place in the solution of resins in alkohol. During metallic dissolutions, on the contrary, a decomposition, either of the acid, or of the water which dilutes it, always takes place; the metal combines with oxygen, and is changed into an oxyd, and a gasseous substance is disengaged; so that in reality none of the substances[Pg 369] employed remain, after the operation, in the same state they were in before. This article is entirely confined to the consideration of solution.

To understand properly what takes place during the solution of salts, it is necessary to know, that, in most of these operations, two distinct effects are complicated together, viz. solution by water, and solution by caloric; and, as the explanation of most of the phenomena of solution depends upon the distinction of these two circumstances, I shall enlarge a little upon their nature.

Nitrat of potash, usually called nitre or saltpetre, contains very little water of cristallization, perhaps even none at all; yet this salt liquifies in a degree of heat very little superior to that of boiling water. This liquifaction cannot therefore be produced by means of the water of cristallization, but in consequence of the salt being very fusible in its nature, and from its passing from the solid to the liquid state of aggregation, when but a little raised above the temperature of boiling water. All salts are in this manner susceptible of being liquified by caloric, but in higher or lower degrees of temperature. Some of these, as the acetites of potash and soda, liquify with a very moderate heat, whilst others, as sulphat of potash, lime, &c. require the strongest fires we are capable of producing. This liquifaction[Pg 370] of salts by caloric produces exactly the same phenomena with the melting of ice; it is accomplished in each salt by a determinate degree of heat, which remains invariably the same during the whole time of the liquifaction. Caloric is employed, and becomes fixed during the melting of the salt, and is, on the contrary, disengaged when the salt coagulates. These are general phenomena which universally occur during the passage of every species of substance from the solid to the fluid state of aggregation, and from fluid to solid.

These phenomena arising from solution by caloric are always less or more conjoined with those which take place during solutions in water. We cannot pour water upon a salt, on purpose to dissolve it, without employing a compound solvent, both water and caloric; hence we may distinguish several different cases of solution, according to the nature and mode of existence of each salt. If, for instance, a salt be difficultly soluble in water, and readily so by caloric, it evidently follows, that this salt will be difficultly soluble in cold water, and considerably in hot water; such is nitrat of potash, and more especially oxygenated muriat of potash. If another salt be little soluble both in water and caloric, the difference of its solubility in cold and warm water will be very inconsiderable; sulphat of lime is of this kind. From these considerations,[Pg 371] it follows, that there is a necessary relation between the following circumstances; the solubility of a salt in cold water, its solubility in boiling water, and the degree of temperature at which the same salt liquifies by caloric, unassisted by water; and that the difference of solubility in hot and cold water is so much greater in proportion to its ready solution in caloric, or in proportion to its susceptibility of liquifying in a low degree of temperature.

The above is a general view of solution; but, for want of particular facts, and sufficiently exact experiments, it is still nothing more than an approximation towards a particular theory. The means of compleating this part of chemical science is extremely simple; we have only to ascertain how much of each salt is dissolved by a certain quantity of water at different degrees of temperature; and as, by the experiments published by Mr de la Place and me, the quantity of caloric contained in a pound of water at each degree of the thermometer is accurately known, it will be very easy to determine, by simple experiments, the proportion of water and caloric required for solution by each salt, what quantity of caloric is absorbed by each at the moment of liquifaction, and how much is disengaged at the moment of cristallization. Hence the reason why salts are more rapidly soluble in hot than in cold water is perfectly evident. In all solutions[Pg 372] of salts caloric is employed; when that is furnished intermediately from the surrounding bodies, it can only arrive slowly to the salt; whereas this is greatly accelerated when the requisite caloric exists ready combined with the water of solution.

In general, the specific gravity of water is augmented by holding salts in solution; but there are some exceptions to the rule. Some time hence, the quantities of radical, of oxygen, and of base, which constitute each neutral salt, the quantity of water and caloric necessary for solution, the increased specific gravity communicated to water, and the figure of the elementary particles of the cristals, will all be accurately known. From these all the circumstances and phenomena of cristallization will be explained, and by these means this part of chemistry will be compleated. Mr Seguin has formed the plan of a thorough investigation of this kind, which he is extremely capable of executing.

The solution of salts in water requires no particular apparatus; small glass phials of different sizes, Pl. II. Fig. 16. and 17. pans of earthern ware, A, Fig. 1. and 2. long-necked matrasses, Fig. 14. and pans or basons of copper or of silver, Fig. 13. and 15. answer very well for these operations.[Pg 373]

SECT. II.
Of Lixiviation.

This is an operation used in chemistry and manufactures for separating substances which are soluble in water from such as are insoluble. The large vat or tub, Pl. II. Fig. 12. having a hole D near its bottom, containing a wooden spiget and fosset or metallic stop-cock DE, is generally used for this purpose. A thin stratum of straw is placed at the bottom of the tub; over this, the substance to be lixiviated is laid and covered by a cloth, then hot or cold water, according to the degree of solubility of the saline matter, is poured on. When the water is supposed to have dissolved all the saline parts, it is let off by the stop-cock; and, as some of the water charged with salt necessarily adheres to the straw and insoluble matters, several fresh quantities of water are poured on. The straw serves to secure a proper passage for the water, and may be compared to the straws or glass rods used in filtrating, to keep the paper from touching the sides of the funnel. The cloth which is laid over the matters under lixiviation prevents the water from making a hollow in[Pg 374] these substances where it is poured on, through which it might escape without acting upon the whole mass.

This operation is less or more imitated in chemical experiments; but as in these, especially with analytical views, greater exactness is required, particular precautions must be employed, so as not to leave any saline or soluble part in the residuum. More water must be employed than in ordinary lixiviations, and the substances ought to be previously stirred up in the water before the clear liquor is drawn off, otherwise the whole mass might not be equally lixiviated, and some parts might even escape altogether from the action of the water. We must likewise employ fresh portions of water in considerable quantity, until it comes off entirely free from salt, which we may ascertain by means of the hydrometer formerly described.

In experiments with small quantities, this operation is conveniently performed in jugs or matrasses of glass, and by filtrating the liquor through paper in a glass funnel. When the substance is in larger quantity, it may be lixiviated in a kettle of boiling water, and filtrated through paper supported by cloth in the wooden frame, Pl. II. Fig. 3. and 4.; and in operations in the large way, the tub already mentioned must be used.[Pg 375]

SECT. III.
Of Evaporation.

This operation is used for separating two substances from each other, of which one at least must be fluid, and whose degrees of volatility are considerably different. By this means we obtain a salt, which has been dissolved in water, in its concrete form; the water, by heating, becomes combined with caloric, which renders it volatile, while the particles of the salt being brought nearer to each other, and within the sphere of their mutual attraction, unite into the solid state.

As it was long thought that the air had great influence upon the quantity of fluid evaporated, it will be proper to point out the errors which this opinion has produced. There certainly is a constant slow evaporation from fluids exposed to the free air; and, though this species of evaporation may be considered in some degree as a solution in air, yet caloric has considerable influence in producing it, as is evident from the refrigeration which always accompanies this process; hence we may consider this gradual evaporation as a compound solution made partly in[Pg 376] air, and partly in caloric. But the evaporation which takes place from a fluid kept continually boiling, is quite different in its nature, and in it the evaporation produced by the action of the air is exceedingly inconsiderable in comparison with that which is occasioned by caloric. This latter species may be termed *vaporization* rather than *evaporation*. This process is not accelerated in proportion to the extent of evaporating surface, but in proportion to the quantities of caloric which combine with the fluid. Too free a current of cold air is often hurtful to this process, as it tends to carry off caloric from the water, and consequently retards its conversion into vapour. Hence there is no inconvenience produced by covering, in a certain degree, the vessels in which liquids are evaporated by continual boiling, provided the covering body be of such a nature as does not strongly draw off the caloric, or, to use an expression of Dr Franklin's, provided it be a bad conductor of heat. In this case, the vapours escape through such opening as is left, and at least as much is evaporated, frequently more than when free access is allowed to the external air.

As during evaporation the fluid carried off by caloric is entirely lost, being sacrificed for the sake of the fixed substances with which it was combined, this process is only employed where the fluid is of small value, as water, for instance.[Pg 377] But, when the fluid is of more consequence, we have recourse to distillation, in which process we preserve both the fixed substance and the volatile fluid. The vessels employed for evaporation are basons or pans of copper, silver, or lead, Pl. II. Fig. 13. and 15. or capsules of glass, porcellain, or stone ware, Pl. II. A, Fig. 1. and 2. Pl. III. Fig. 3 and 4. The best utensils for this purpose are made of the bottoms of glass retorts and matrasses, as their equal thinness renders them more fit than any other kind of glass vessel for bearing a brisk fire and sudden alterations of heat and cold without breaking.

As the method of cutting these glass vessels is no where described in books, I shall here give a description of it, that they may be made by chemists for themselves out of spoiled retorts, matrasses, and recipients, at a much cheaper rate than any which can be procured from glass manufacturers. The instrument, Pl. III. Fig. 5. consisting of an iron ring AC, fixed to the rod AB, having a wooden handle D, is employed as follows: Make the ring red hot in the fire, and put it upon the matrass G, Fig. 6. which is to be cut; when the glass is sufficiently heated, throw on a little cold water, and it will generally break exactly at the circular line heated by the ring.

Small flasks or phials of thin glass are exceeding good vessels for evaporating small quantities[Pg 378] of fluid; they are very cheap, and stand the fire remarkably. One or more of these may be placed upon a second grate above the furnace, Pl. III. Fig. 2. where they will only experience a gentle heat. By this means a great number of experiments may be carried on at one time. A glass retort, placed in a sand bath, and covered with a dome of baked earth, Pl. III. Fig. 1. answers pretty well for evaporations; but in this way it is always considerably slower, and is even liable to accidents; as the sand heats unequally, and the glass cannot dilate in the same unequal manner, the retort is very liable to break. Sometimes the sand serves exactly the office of the iron ring formerly mentioned; for, if a single drop of vapour, condensed into liquid, happens to fall upon the heated part of the vessel, it breaks circularly at that place. When a very intense fire is necessary, earthen crucibles may be used; but we generally use the word *evaporation* to express what is produced by the temperature of boiling water, or not much higher.[Pg 379]

SECT. IV.
Of Cristallization.

In this process the integrant parts of a solid body, separated from each other by the intervention of a fluid, are made to exert the mutual attraction of aggregation, so as to coalesce and reproduce a solid mass. When the particles of a body are only separated by caloric, and the substance is thereby retained in the liquid state, all that is necessary for making it cristallize, is to remove a part of the caloric which is lodged between its particles, or, in other words, to cool it. If this refrigeration be slow, and the body be at the same time left at rest, its particles assume a regular arrangement, and cristallization, properly so called, takes place; but, if the refrigeration is made rapidly, or if the liquor be agitated at the moment of its passage to the concrete state, the cristallization is irregular and confused.

The same phenomena occur with watery solutions, or rather in those made partly in water, and partly by caloric. So long as there remains a sufficiency of water and caloric to keep the particles of the body asunder beyond the sphere[Pg 380] of their mutual attraction, the salt remains in the fluid state; but, whenever either caloric or water is not present in sufficient quantity, and the attraction of the particles for each other becomes superior to the power which keeps them asunder, the salt recovers its concrete form, and the cristals produced are the more regular in proportion as the evaporation has been slower and more tranquilly performed.

All the phenomena we formerly mentioned as taking place during the solution of salts, occur in a contrary sense during their cristallization. Caloric is disengaged at the instant of their assuming the solid state, which furnishes an additional proof of salt being held in solution by the compound action of water and caloric. Hence, to cause salts to cristallize which readily liquify by means of caloric, it is not sufficient to carry off the water which held them in solution, but the caloric united to them must likewise be removed. Nitrat of potash, oxygenated muriat of potash, alum, sulphat of soda, &c. are examples of this circumstance, as, to make these salts cristallize, refrigeration must be added to evaporation. Such salts, on the contrary, as require little caloric for being kept in solution, and which, from that circumstance, are nearly equally soluble in cold and warm water, are cristallizable by simply carrying off the water which holds them in solution, and[Pg 381] even recover their solid state in boiling water; such are sulphat of lime, muriat of potash and of soda, and several others.

The art of refining saltpetre depends upon these properties of salts, and upon their different degrees of solubility in hot and cold water. This salt, as produced in the manufactories by the first operation, is composed of many different salts; some are deliquescent, and not susceptible of being cristallized, such as the nitrat and muriat of lime; others are almost equally soluble in hot and cold water, as the muriats of potash and of soda; and, lastly, the saltpetre, or nitrat of potash, is greatly more soluble in hot than it is in cold water. The operation is begun, by pouring upon this mixture of salts as much water as will hold even the least soluble, the muriats of soda and of potash, in solution; so long as it is hot, this quantity readily dissolves all the saltpetre, but, upon cooling, the greater part of this salt cristallizes, leaving about a sixth part remaining dissolved, and mixed with the nitrat of lime and the two muriats. The nitre obtained by this process is still somewhat impregnated with other salts, because it has been cristallized from water in which these abound: It is completely purified from these by a second solution in a small quantity of boiling water, and second cristallization. The water remaining after these cristallizations of nitre is still loaded with a mixture[Pg 382] of saltpetre, and other salts; by farther evaporation, crude saltpetre, or rough-petre, as the workmen call it, is procured from it, and this is purified by two fresh solutions and cristallizations.

The deliquescent earthy salts which do not contain the nitric acid are rejected in this manufacture; but those which consist of that acid neutralized by an earthy base are dissolved in water, the earth is precipitated by means of potash, and allowed to subside; the clear liquor is then decanted, evaporated, and allowed to cristallize. The above management for refining saltpetre may serve as a general rule for separating salts from each other which happen to be mixed together. The nature of each must be considered, the proportion in which each dissolves in given quantities of water, and the different solubility of each in hot and cold water. If to these we add the property which some salts possess, of being soluble in alkohol, or in a mixture of alkohol and water, we have many resources for separating salts from each other by means of cristallization, though it must be allowed that it is extremely difficult to render this separation perfectly complete.

The vessels used for cristallization are pans of earthen ware, A, Pl. II. Fig. 1. and 2. and large flat dishes, Pl. III. Fig. 7. When a saline solution is to be exposed to a slow evaporation[Pg 383] in the heat of the atmosphere, with free access of air, vessels of some depth, Pl. III. Fig. 3. must be employed, that there may be a considerable body of liquid; by this means the cristals produced are of considerable size, and remarkably regular in their figure.

Every species of salt cristallizes in a peculiar form, and even each salt varies in the form of its cristals according to circumstances, which take place during cristallization. We must not from thence conclude that the saline particles of each species are indeterminate in their figures: The primative particles of all bodies, especially of salts, are perfectly constant in their specific forms; but the cristals which form in our experiments are composed of congeries of minute particles, which, though perfectly equal in size and shape, may assume very dissimilar arrangements, and consequently produce a vast variety of regular forms, which have not the smallest apparent resemblance to each other, nor to the original cristal. This subject has been very ably treated by the Abbé Haüy, in several memoirs presented to the Academy, and in his work upon the structure of cristals: It is only necessary to extend generally to the class of salts the principles he has particularly applied to some cristalized stones.[Pg 384]

SECT. V.
Of Simple Distillation.

As distillation has two distinct objects to accomplish, it is divisible into simple and compound; and, in this section, I mean to confine myself entirely to the former. When two bodies, of which one is more volatile than the other, or has more affinity to caloric, are submitted to distillation, our intention is to separate them from each other: The more volatile substance assumes the form of gas, and is afterwards condensed by refrigeration in proper vessels. In this case distillation, like evaporation, becomes a species of mechanical operation, which separates two substances from each other without decomposing or altering the nature of either. In evaporation, our only object is to preserve the fixed body, without paying any regard to the volatile matter; whereas, in distillation, our principal attention is generally paid to the volatile substance, unless when we intend to preserve both the one and the other. Hence, simple distillation is nothing more than evaporation produced in close vessels.

The most simple distilling vessel is a species of bottle or matrass, A, Pl. III. Fig. 8. which has[Pg 385] been bent from its original form BC to BD, and which is then called a retort; when used, it is placed either in a reverberatory furnace, Pl. XIII. Fig. 2. or in a sand bath under a dome of baked earth, Pl. III. Fig. 1. To receive and condense the products, we adapt a recipient, E, Pl. III. Fig. 9. which is luted to the retort. Sometimes, more especially in pharmaceutical operations, the glass or stone ware cucurbit, A, with its capital B, Pl. III. Fig. 12, or the glass alembic and capital, Fig. 13. of one piece, is employed. This latter is managed by means of a tubulated opening T, fitted with a ground stopper of cristal; the capital, both of the cucurbit and alembic, has a furrow or trench, *r r*, intended for conveying the condensed liquor into the beak RS, by which it runs out. As, in almost all distillations, expansive vapours are produced, which might burst the vessels employed, we are under the necessity of having a small hole, T, Fig. 9. in the balloon or recipient, through which these may find vent; hence, in this way of distilling, all the products which are permanently aëriform are entirely lost, and even such as difficultly lose that state have not sufficient space to condense in the balloon: This apparatus is not, therefore, proper for experiments of investigation, and can only be admitted in the ordinary operations of the laboratory or in pharmacy. In the article appropriated for compound[Pg 386] distillation, I shall explain the various methods which have been contrived for preserving the whole products from bodies in this process.

As glass or earthen vessels are very brittle, and do not readily bear sudden alterations of heat and cold, every well regulated laboratory ought to have one or more alembics of metal for distilling water, spirituous liquors, essential oils, &c. This apparatus consists of a cucurbit and capital of tinned copper or brass, Pl. III. Fig. 15. and 16. which, when judged proper, may be placed in the water bath, D, Fig. 17. In distillations, especially of spiritous liquors, the capital must be furnished with a refrigeratory, SS, Fig. 16. kept continually filled with cold water; when the water becomes heated, it is let off by the stop-cock, R, and renewed with a fresh supply of cold water. As the fluid distilled is converted into gas by means of caloric furnished by the fire of the furnace, it is evident that it could not condense, and, consequently, that no distillation, properly speaking, could take place, unless it is made to deposit in the capital all the caloric received in the cucurbit; with this view, the sides of the capital must always be preserved at a lower temperature than is necessary for keeping the distilling substance in the state of gas, and the water in the refrigeratory is intended for this purpose.[Pg 387] Water is converted into gas by the temperature of 80° (212°), alkohol by 67° (182.75°), ether by 32° (104°); hence these substances cannot be distilled, or, rather, they will fly off in the state of gas, unless the temperature of the refrigetory be kept under these respective degrees.

In the distillation of spiritous, and other expansive liquors, the above described refrigetory is not sufficient for condensing all the vapours which arise; in this case, therefore, instead of receiving the distilled liquor immediately from the beak, TU, of the capital into a recipient, a worm is interposed between them. This instrument is represented Pl. III. Fig. 18. contained in a worm tub of tinned copper, it consists of a metallic tube bent into a considerable number of spiral revolutions. The vessel which contains the worm is kept full of cold water, which is renewed as it grows warm. This contrivance is employed in all distilleries of spirits, without the intervention of a capital and refrigeratory, properly so called. The one represented in the plate is furnished with two worms, one of them being particularly appropriated to distillations of odoriferous substances.

In some simple distillations it is necessary to interpose an adopter between the retort and receiver, as shown Pl. III. Fig, 11. This may[Pg 388] serve two different purposes, either to separate two products of different degrees of volatility, or to remove the receiver to a greater distance from the furnace, that it may be less heated. But these, and several other more complicated instruments of ancient contrivance, are far from producing the accuracy requisite in modern chemistry, as will be readily perceived when I come to treat of compound distillation.

SECT. VI.

Of Sublimation.
This term is applied to the distillation of substances which condense in a concrete or solid form, such as the sublimation of sulphur, and of muriat of ammoniac, or sal ammoniac. These operations may be conveniently performed in the ordinary distilling vessels already described, though, in the sublimation of sulphur, a species of vessels, named Alludels, have been usually employed. These are vessels of stone or porcelain ware, which adjust to each other over a cucurbit containing the sulphur to be sublimed. One of the best subliming vessels, for substances which are not very volatile, is a flask,[Pg 389] or phial of glass, sunk about two thirds into a sand bath; but in this way we are apt to lose a part of the products. When these are wished to be entirely preserved, we must have recourse to the pneumato-chemical distilling apparatus, to be described in the following chapter.

[Pg 390]

CHAP. VI.
Of Pneumato-chemical Distillations, Metallic Dissolutions, and some other operations which require very complicated instruments.
SECT. I.
Of Compound and Pneumato-chemical Distillations.

In the preceding chapter, I have only treated of distillation as a simple operation, by which two substances, differing in degrees of volatility, may be separated from each other; but distillation often actually decomposes the substances submitted to its action, and becomes one of the most complicated operations in chemistry. In every distillation, the substance distilled must be brought to the state of gas, in the cucurbit or retort, by combination with caloric: In simple distillation, this caloric is given out in the refrigeratory or in the worm, and the substance again recovers its liquid or solid form, but the substances submitted to compound distillation[Pg 391] are absolutely decompounded; one part, as for instance the charcoal they contain, remains fixed in the retort, and all the rest of the elements are reduced to gasses of different kinds. Some of these are susceptible of being condensed, and of recovering their solid or liquid forms, whilst others are permanently aëriform; one part of these are absorbable by water, some by the alkalies, and others are not susceptible of being absorbed at all. An ordinary distilling apparatus, such as has been described in the preceding chapter, is quite insufficient for retaining or for separating these diversified products, and we are obliged to have recourse, for this purpose, to methods of a more complicated nature.

The apparatus I am about to describe is calculated for the most complicated distillations, and may be simplified according to circumstances. It consists of a tubulated glass retort A, Pl. IV. Fig. 1. having its beak fitted to a tubulated balloon or recipient BC; to the upper orifice D of the balloon a bent tube DE*fg* is adjusted, which, at its other extremity *g*, is plunged into the liquor contained in the bottle L, with three necks *xxx*. Three other similar bottles are connected with this first one, by means of three similar bent tubes disposed in the same manner; and the farthest neck of the last bottle is connected with a jar in a pneumato-chemical apparatus, by means of a bent[Pg 392] tube[60]. A determinate weight of distilled water is usually put into the first bottle, and the other three have each a solution of caustic potash in water. The weight of all these bottles, and of the water and alkaline solution they contain, must be accurately ascertained. Every thing being thus disposed, the junctures between the retort and recipient, and of the tube D of the latter, must be luted with fat lute, covered over with slips of linen, spread with lime and white of egg; all the other junctures are to be secured by a lute made of wax and rosin melted together.

When all these dispositions are completed, and when, by means of heat applied to the retort A, the substance it contains becomes decomposed, it is evident that the least volatile products must condense or sublime in the beak or neck of the retort itself, where most of the concrete substances will fix themselves. The more volatile substances, as the lighter oils, ammoniac, and several others, will condense in the recipient GC, whilst the gasses, which are not susceptible of condensation by cold, will pass on by the tubes, and boil up through the liquors in the several bottles. Such as are absorbable[Pg 393] by water will remain in the first bottle, and those which caustic alkali can absorb will remain in the others; whilst such gasses as are not susceptible of absorption, either by water or alkalies, will escape by the tube RM, at the end of which they may be received into jars in a pneumato-chemical apparatus. The charcoal and fixed earth, &c. which form the substance or residuum, anciently called *caput mortuum*, remain behind in the retort.

In this manner of operating, we have always a very material proof of the accuracy of the analysis, as the whole weights of the products taken together, after the process is finished, must be exactly equal to the weight of the original substance submitted to distillation. Hence, for instance, if we have operated upon eight ounces of starch or gum arabic, the weight of the charry residuum in the retort, together with that of all the products gathered in its neck and the balloon, and of all the gas received into the jars by the tube RM added to the additional weight acquired by the bottles, must, when taken together, be exactly eight ounces. If the product be less or more, it proceeds from error, and the experiment must be repeated until a satisfactory result be procured, which ought not to differ more than six or eight grains in the pound from the weight of the substance submitted to experiment.[Pg 394]

In experiments of this kind, I for a long time met with an almost insurmountable difficulty, which must at last have obliged me to desist altogether, but for a very simple method of avoiding it, pointed out to me by Mr Hassenfratz. The smallest diminution in the heat of the furnace, and many other circumstances inseparable from this kind of experiments, cause frequent reabsorptions of gas; the water in the cistern of the pneumato-chemical apparatus rushes into the last bottle through the tube RM, the same circumstance happens from one bottle into another, and the fluid is often forced even into the recipient C. This accident is prevented by using bottles having three necks, as represented in the plate, into one of which, in each bottle, a capillary glass-tube St, st, st, st, is adapted, so as to have its lower extremity t immersed in the liquor. If any absorption takes place, either in the retort, or in any of the bottles, a sufficient quantity of external air enters, by means of these tubes, to fill up the void; and we get rid of the inconvenience at the price of having a small mixture of common air with the products of the experiment, which is thereby prevented from failing altogether. Though these tubes admit the external air, they cannot permit any of the gasseous substances to escape, as they are always shut below by the water of the bottles.[Pg 395]

It is evident that, in the course of experiments with this apparatus, the liquor of the bottles must rise in these tubes in proportion to the pressure sustained by the gas or air contained in the bottles; and this pressure is determined by the height and gravity of the column of fluid contained in all the subsequent bottles. If we suppose that each bottle contains three inches of fluid, and that there are three inches of water in the cistern of the connected apparatus above the orifice of the tube RM, and allowing the gravity of the fluids to be only equal to that of water, it follows that the air in the first bottle must sustain a pressure equal to twelve inches of water; the water must therefore rise twelve inches in the tube S, connected with the first bottle, nine inches in that belonging to the second, six inches in the third, and three in the last; wherefore these tubes must be made somewhat more than twelve, nine, six, and three inches long respectively, allowance being made for oscillatory motions, which often take place in the liquids. It is sometimes necessary to introduce a similar tube between the retort and recipient; and, as the tube is not immersed in fluid at its lower extremity, until some has collected in the progress of the distillation, its upper end must be shut at first with a little lute, so as to be opened according to necessity, or after[Pg 396] there is sufficient liquid in the recipient to secure its lower extremity.

This apparatus cannot be used in very accurate experiments, when the substances intended to be operated upon have a very rapid action upon each other, or when one of them can only be introduced in small successive portions, as in such as produce violent effervescence when mixed together. In such cases, we employ a tubulated retort A, Pl. VII. Fig. 1. into which one of the substances is introduced, preferring always the solid body, if any such is to be treated, we then lute to the opening of the retort a bent tube BCDA, terminating at its upper extremity B in a funnel, and at its other end A in a capillary opening. The fluid material of the experiment is poured into the retort by means of this funnel, which must be made of such a length, from B to C, that the column of liquid introduced may counterbalance the resistance produced by the liquors contained in all the bottles, Pl. IV. Fig. 1.

Those who have not been accustomed to use the above described distilling apparatus may perhaps be startled at the great number of openings which require luting, and the time necessary for making all the previous preparations in experiments of this kind. It is very true that, if we take into account all the necessary weighings of materials and products, both before and[Pg 397] after the experiments, these preparatory and succeeding steps require much more time and attention than the experiment itself. But, when the experiment succeeds properly, we are well rewarded for all the time and trouble bestowed, as by one process carried on in this accurate manner much more just and extensive knowledge is acquired of the nature of the vegetable or animal substance thus submitted to investigation, than by many weeks assiduous labour in the ordinary method of proceeding.

When in want of bottles with three orifices, those with two may be used; it is even possible to introduce all the three tubes at one opening, so as to employ ordinary wide-mouthed bottles, provided the opening be sufficiently large. In this case we must carefully fit the bottles with corks very accurately cut, and boiled in a mixture of oil, wax, and turpentine. These corks are pierced with the necessary holes for receiving the tubes by means of a round file, as in Pl. IV. Fig. 8.[Pg 398]

SECT. II.
Of Metallic Dissolutions.

I have already pointed out the difference between solution of salts in water and metallic dissolutions. The former requires no particular vessels, whereas the latter requires very complicated vessels of late invention, that we may not lose any of the products of the experiment, and may thereby procure truly conclusive results of the phenomena which occur. The metals, in general, dissolve in acids with effervescence, which is only a motion excited in the solvent by the disengagement of a great number of bubbles of air or aëriform fluid, which proceed from the surface of the metal, and break at the surface of the liquid.

Mr Cavendish and Dr Priestley were the first inventors of a proper apparatus for collecting these elastic fluids. That of Dr Priestley is extremely simple, and consists of a bottle A, Pl. VII. Fig. 2. with its cork B, through which passes the bent glass tube BC, which is engaged under a jar filled with water in the pneumato-chemical apparatus, or simply in a bason full of water. The metal is first introduced into the[Pg 399]bottle, the acid is then poured over it, and the bottle is instantly closed with its cork and tube, as represented in the plate. But this apparatus has its inconveniencies. When the acid is much concentrated, or the metal much divided, the effervescence begins before we have time to cork the bottle properly, and some gas escapes, by which we are prevented from ascertaining the quantity disengaged with rigorous exactness. In the next place, when we are obliged to employ heat, or when heat is produced by the process, a part of the acid distills, and mixes with the water of the pneumato-chemical apparatus, by which means we are deceived in our calculation of the quantity of acid decomposed. Besides these, the water in the cistern of the apparatus absorbs all the gas produced which is susceptible of absorption, and renders it impossible to collect these without loss.

To remedy these inconveniencies, I at first used a bottle with two necks, Pl. VII. Fig. 3. into one of which the glass funnel BC is luted so as to prevent any air escaping; a glass rod DE is fitted with emery to the funnel, so as to serve the purpose of a stopper. When it is used, the matter to be dissolved is first introduced into the bottle, and the acid is then permitted to pass in as slowly as we please, by raising the glass rod gently as often as is necessary until saturation is produced.[Pg 400]

Another method has been since employed, which serves the same purpose, and is preferable to the last described in some instances. This consists in adapting to one of the mouths of the bottle A, Pl. VII. Fig. 4. a bent tube DEFG, having a capillary opening at D, and ending in a funnel at G. This tube is securely luted to the mouth C of the bottle. When any liquid is poured into the funnel, it falls down to F; and, if a sufficient quantity be added, it passes by the curvature E, and falls slowly into the bottle, so long as fresh liquor is supplied at the funnel. The liquor can never be forced out of the tube, and no gas can escape through it, because the weight of the liquid serves the purpose of an accurate cork.

To prevent any distillation of acid, especially in dissolutions accompanied with heat, this tube is adapted to the retort A, Pl. VII. Fig. 1. and a small tubulated recipient, M, is applied, in which any liquor which may distill is condensed. On purpose to separate any gas that is absorbable by water, we add the double necked bottle L, half filled with a solution of caustic potash; the alkali absorbs any carbonic acid gas, and usually only one or two other gasses pass into the jar of the connected pneumato-chemical apparatus through the tube NO. In the first chapter of this third part we have directed how these are to be separated and examined.[Pg 401] If one bottle of alkaline solution be not thought sufficient, two, three, or more, may be added.

SECT. III.
Apparatus necessary in Experiments upon Vinous and Putrefactive Fermentations.

For these operations a peculiar apparatus, especially intended for this kind of experiment, is requisite. The one I am about to describe is finally adopted, as the best calculated for the purpose, after numerous corrections and improvements. It consists of a large matrass, A, Pl. X. fig. 1. holding about twelve pints, with a cap of brass a b, strongly cemented to its mouth, and into which is screwed a bent tube c d, furnished with a stop-cock e. To this tube is joined the glass recipient B, having three openings, one of which communicates with the bottle C, placed below it. To the posterior opening of this recipient is fitted a glass tube g h i, cemented at g and i to collets of brass, and intended to contain a very deliquescent concrete neutral salt, such as nitrat or muriat of lime, acetite of potash, &c. This tube communicates with two bottles D and E, filled to x and y with a solution of caustic potash.[Pg 402]

All the parts of this machine are joined together by accurate screws, and the touching parts have greased leather interposed, to prevent any passage of air. Each piece is likewise furnished with two stop-cocks, by which its two extremities may be closed, so that we can weigh each separately at any period of the operation.

The fermentable matter, such as sugar, with a proper quantity of yeast, and diluted with water, is put into the matrass. Sometimes, when the fermentation is too rapid, a considerable quantity of froth is produced, which not only fills the neck of the matrass, but passes into the recipient, and from thence runs down into the bottle C. On purpose to collect this scum and must, and to prevent it from reaching the tube filled with deliquescent salts, the recipient and connected bottle are made of considerable capacity.

In the vinous fermentation, only carbonic acid gas is disengaged, carrying with it a small proportion of water in solution. A great part of this water is deposited in passing through the tube g h i, which is filled with a deliquescent salt in gross powder, and the quantity is ascertained by the augmentation of the weight of the salt. The carbonic acid gas bubbles up through the alkaline solution in the bottle D, to which it is conveyed by the tube k l m. Any small portion which may not be absorbed by this[Pg 403] first bottle is secured by the solution in the second bottle E, so that nothing, in general, passes into the jar F, except the common air contained in the vessels at the commencement of the experiment.

The same apparatus answers extremely well for experiments upon the putrefactive fermentation; but, in this case, a considerable quantity of hydrogen gas is disengaged through the tube q r s t u, by which it is conveyed into the jar F; and, as this disengagement is very rapid, especially in summer, the jar must be frequently changed. These putrefactive fermentations require constant attendance from the above circumstance, whereas the vinous fermentation hardly needs any. By means of this apparatus we can ascertain, with great precision, the weights of the substances submitted to fermentation, and of the liquid and aëriform products which are disengaged. What has been already said in Part I. Chap. XIII. upon the products of the vinous fermentation, may be consulted.[Pg 404]

SECT. IV.
Apparatus for the Decomposition of Water.

Having already given an account, in the first part of this work, of the experiments relative to the decomposition of water, I shall avoid any unnecessary repetitions, and only give a few summary observations upon the subject in this section. The principal substances which have the power of decomposing water are iron and charcoal; for which purpose, they require to be made red hot, otherwise the water is only reduced into vapours, and condenses afterwards by refrigeration, without sustaining the smallest alteration. In a red heat, on the contrary, iron or charcoal carry off the oxygen from its union with hydrogen; in the first case, black oxyd of iron is produced, and the hydrogen is disengaged pure in form of gas; in the other case, carbonic acid gas is formed, which disengages, mixed with the hydrogen gas; and this latter is commonly carbonated, or holds charcoal in solution.

A musket barrel, without its breach pin, answers exceedingly well for the decomposition of water, by means of iron, and one should be[Pg 405] chosen of considerable length, and pretty strong. When too short, so as to run the risk of heating the lute too much, a tube of copper is to be strongly soldered to one end. The barrel is placed in a long furnace, CDEF, Pl. VII. Fig. 11. so as to have a few degrees of inclination from E to F; a glass retort A, is luted to the upper extremity E, which contains water, and is placed upon the furnace VVXX. The lower extremity F is luted to a worm SS, which is connected with the tubulated bottle H, in which any water distilled without decomposition, during the operation, collects, and the disengaged gas is carried by the tube KK to jars in a pneumato-chemical apparatus. Instead of the retort a funnel may be employed, having its lower part shut by a stop-cock, through which the water is allowed to drop gradually into the gun-barrel. Immediately upon getting into contact with the heated part of the iron, the water is converted into steam, and the experiment proceeds in the same manner as if it were furnished in vapours from the retort.

In the experiment made by Mr Meusnier and me before a committee of the Academy, we used every precaution to obtain the greatest possible precision in the result of our experiment, having even exhausted all the vessels employed before we began, so that the hydrogen gas obtained might be free from any mixture[Pg 406] of azotic gas. The results of that experiment will hereafter be given at large in a particular memoir.

In numerous experiments, we are obliged to use tubes of glass, porcelain, or copper, instead of gun-barrels; but glass has the disadvantage of being easily melted and flattened, if the heat be in the smallest degree raised too high; and porcelain is mostly full of small minute pores, through which the gas escapes, especially when compressed by a column of water. For these reasons I procured a tube of brass, which Mr de la Briche got cast and bored out of the solid for me at Strasburg, under his own inspection. This tube is extremely convenient for decomposing alkohol, which resolves into charcoal, carbonic acid gas, and hydrogen gas; it may likewise be used with the same advantage for decomposing water by means of charcoal, and in a great number of experiments of this nature.[Pg 407]

FOOTNOTES:

[60]The representation of this apparatus, Pl. IV. Fig. 1. will convey a much better idea of its disposition than can possibly be given by the most laboured description.—E.

CHAP. VII.
Of the Composition and Application of Lutes.

The necessity of properly securing the junctures of chemical vessels to prevent the escape of any of the products of experiments, must be sufficiently apparent; for this purpose lutes are employed, which ought to be of such a nature as to be equally impenetrable to the most subtile substances, as glass itself, through which only caloric can escape.

This first object of lutes is very well accomplished by bees wax, melted with about an eighth part of turpentine. This lute is very easily managed, sticks very closely to glass, and is very difficultly penetrable; it may be rendered more consistent, and less or more hard or pliable, by adding different kinds of resinous matters. Though this species of lute answers extremely well for retaining gasses and vapours, there are many chemical experiments which produce considerable heat, by which this lute becomes liquified, and consequently the expansive vapours must very readily force through and escape.[Pg 408]

For such cases, the following fat lute is the best hitherto discovered, though not without its disadvantages, which shall be pointed out. Take very pure and dry unbaked clay, reduced to a very fine powder, put this into a brass mortar, and beat it for several hours with a heavy iron pestle, dropping in slowly some boiled lintseed oil; this is oil which has been oxygenated, and has acquired a drying quality, by being boiled with litharge. This lute is more tenacious, and applies better, if amber varnish be used instead of the above oil. To make this varnish, melt some yellow amber in an iron laddle, by which operation it loses a part of its succinic acid, and essential oil, and mix it with lintseed oil. Though the lute prepared with this varnish is better than that made with boiled oil, yet, as its additional expence is hardly compensated by its superior quality, it is seldom used.

The above fat lute is capable of sustaining a very violent degree of heat, is impenetrable by acids and spiritous liquors, and adheres exceedingly well to metals, stone ware, or glass, providing they have been previously rendered perfectly dry. But if, unfortunately, any of the liquor in the course of an experiment gets through, either between the glass and the lute, or between the layers of the lute itself, so as to moisten the part, it is extremely difficult to close[Pg 409] the opening. This is the chief inconvenience which attends the use of fat lute, and perhaps the only one it is subject to. As it is apt to soften by heat, we must surround all the junctures with slips of wet bladder applied over the luting, and fixed on by pack-thread tied round both above and below the joint; the bladder, and consequently the lute below, must be farther secured by a number of turns of pack-thread all over it. By these precautions, we are free from every danger of accident; and the junctures secured in this manner may be considered, in experiments, as hermetically sealed.

It frequently happens that the figure of the junctures prevents the application of ligatures, which is the case with the three-necked bottles formerly described; and it even requires great address to apply the twine without shaking the apparatus; so that, where a number of junctures require luting, we are apt to displace several while securing one. In these cases, we may substitute slips of linen, spread with white of egg and lime mixed together, instead of the wet bladder. These are applied while still moist, and very speedily dry and acquire considerable hardness. Strong glue dissolved in water may answer instead of white of egg. These fillets are usefully applied likewise over junctures luted together with wax and rosin.[Pg 410]

Before applying a lute, all the junctures of the vessels must be accurately and firmly fitted to each other, so as not to admit of being moved. If the beak of a retort is to be luted to the neck of a recipient, they ought to fit pretty accurately; otherwise we must fix them, by introducing short pieces of soft wood or of cork. If the disproportion between the two be very considerable, we must employ a cork which fits the neck of the recipient, having a circular hole of proper dimensions to admit the beak of the retort. The same precaution is necessary in adapting bent tubes to the necks of bottles in the apparatus represented Pl. IV. Fig. 1. and others of a similar nature. Each mouth of each bottle must be fitted with a cork, having a hole made with a round file of a proper size for containing the tube. And, when one mouth is intended to admit two or more tubes, which frequently happens when we have not a sufficient number of bottles with two or three necks, we must use a cork with two or three holes, Pl. IV. Fig. 8.

When the whole apparatus is thus solidly joined, so that no part can play upon another, we begin to lute. The lute is softened by kneading and rolling it between the fingers, with the assistance of heat, if necessary. It is rolled into little cylindrical pieces, and applied to the junctures, taking great care to make it[Pg 411] apply close, and adhere firmly, in every part; a second roll is applied over the first, so as to pass it on each side, and so on till each juncture be sufficiently covered; after this, the slips of bladder, or of linen, as above directed, must be carefully applied over all. Though this operation may appear extremely simple, yet it requires peculiar delicacy and management; great care must be taken not to disturb one juncture whilst luting another, and more especially when applying the fillets and ligatures.

Before beginning any experiment, the closeness of the luting ought always to be previously tried, either by slightly heating the retort A, Pl. IV. Fig. 1, or by blowing in a little air by some of the perpendicular tubes S s s s; the alteration of pressure causes a change in the level of the liquid in these tubes. If the apparatus be accurately luted, this alteration of level will be permanent; whereas, if there be the smallest, opening in any of the junctures, the liquid will very soon recover its former level. It must always be remembered, that the whole success of experiments in modern chemistry depends upon the exactness of this operation, which therefore requires the utmost patience, and most attentive accuracy.

It would be of infinite service to enable chemists, especially those who are engaged in pneumatic processes, to dispense with the use of lutes,[Pg 412] or at least to diminish the number necessary in complicated instruments. I once thought of having my apparatus constructed so as to unite in all its parts by fitting with emery, in the way of bottles with cristal stoppers; but the execution of this plan was extremely difficult. I have since thought it preferable to substitute columns of a few lines of mercury in place of lutes, and have got an apparatus constructed upon this principle, which appears capable of very convenient application in a great number of circumstances.

It consists of a double necked bottle A, Pl. XII. Fig. 12.; the interior neck *bc* communicates with the inside of the bottle, and the exterior neck or rim *de* leaves an interval between the two necks, forming a deep gutter intended to contain the mercury. The cap or lid of glass B enters this gutter, and is properly fitted to it, having notches in its lower edge for the passage of the tubes which convey the gas. These tubes, instead of entering directly into the bottles as in the ordinary apparatus, have a double bend for making them enter the gutter, as represented in Fig. 13. and for making them fit the notches of the cap B; they rise again from the gutter to enter the inside of the bottle over the border of the inner mouth. When the tubes are disposed in their proper places, and the cap firmly fitted on, the gutter is filled with[Pg 413] mercury, by which means the bottle is completely excluded from any communication, excepting through the tubes. This apparatus may be very convenient in many operations in which the substances employed have no action upon Mercury. Pl. XII. Fig. 14. represents an apparatus upon this principle properly fitted together.

Mr Seguin, to whose active and intelligent assistance I have been very frequently much indebted, has bespoken for me, at the glass-houses, some retorts hermetically united to their recipients, by which luting will be altogether unnecessary.

[Pg 414]

CHAP. VIII.
Of Operations upon Combustion and Deflagration.
SECT. I.
Of Combustion in general.

Combustion, according to what has been already said in the First Part of this Work, is the decomposition of oxygen gas produced by a combustible body. The oxygen which forms the base of this gas is absorbed by, and enters into, combination with the burning body, while the caloric and light are set free. Every combustion, therefore, necessarily supposes oxygenation; whereas, on the contrary, every oxygenation does not necessarily imply concomitant combustion; because combustion, properly so called, cannot take place without disengagement of caloric and light. Before combustion can take place, it is necessary that the base of oxygen gas should have greater affinity to the combustible body than it has to caloric;[Pg 415] and this elective attraction, to use Bergman's expression, can only take place at a certain degree of temperature, which is different for each combustible substance; hence the necessity of giving a first motion or beginning to every combustion by the approach of a heated body. This necessity of heating any body we mean to burn depends upon certain considerations, which have not hitherto been attended to by any natural philosopher, for which reason I shall enlarge a little upon the subject in this place.

Nature is at present in a state of equilibrium, which cannot have been attained until all the spontaneous combustions or oxygenations possible in the ordinary degrees of temperature had taken place. Hence, no new combustions or oxygenations can happen without destroying this equilibrium, and raising the combustible substances to a superior degree of temperature. To illustrate this abstract view of the matter by example: Let us suppose the usual temperature of the earth a little changed, and that it is raised only to the degree of boiling water; it is evident, that, in this case, phosphorus, which is combustible in a considerably lower degree of temperature, would no longer exist in nature in its pure and simple state, but would always be procured in its acid or oxygenated state, and its radical would become one of the substances unknown[Pg 416] to chemistry. By gradually increasing the temperature of the earth the same circumstance would successively happen to all the bodies capable of combustion; and, at last, every possible combustion having taken place, there would no longer exist any combustible body whatever, as every substance susceptible of that operation would be oxygenated, and consequently incombustible.

There cannot therefore exist, so far as relates to us, any combustible body, except such as are incombustible in the ordinary temperatures of the earth; or, what is the same thing, in other words, that it is essential to the nature of every combustible body not to possess the property of combustion, unless heated, or raised to the degree of temperature at which its combustion naturally takes place. When this degree is once produced, combustion commences, and the caloric which is disengaged by the decomposition of the oxygen gas keeps up the temperature necessary for continuing combustion. When this is not the case, that is, when the disengaged caloric is insufficient for keeping up the necessary temperature, the combustion ceases: This circumstance is expressed in common language by saying, that a body burns ill, or with difficulty.

Although combustion possesses some circumstances in common with distillation, especially[Pg 417] with the compound kind of that operation, they differ in a very material point. In distillation there is a separation of one part of the elements of the substance from each other, and a combination of these, in a new order, occasioned by the affinities which take place in the increased temperature produced during distillation: This likewise happens in combustion, but with this farther circumstance, that a new element, not originally in the body, is brought into action; oxygen is added to the substance submitted to the operation, and caloric is disengaged.

The necessity of employing oxygen in the state of gas in all experiments with combustion, and the rigorous determination of the quantities employed, render this kind of operations peculiarly troublesome. As almost all the products of combustion are disengaged in the state of gas, it is still more difficult to retain them than even those furnished during compound distillation; hence this precaution was entirely neglected by the ancient chemists; and this set of experiments exclusively belong to modern chemistry.

Having thus pointed out, in a general way, the objects to be had in view in experiments upon combustion, I proceed, in the following sections of this chapter, to describe the different instruments I have used with this view. The following arrangement is formed, not upon the[Pg 418] nature of the combustible bodies, but upon that of the instruments necessary for combustion.

SECT. II.
Of the Combustion of Phosphorus.

In these combustions we begin by filling a jar, capable at least of holding six pints, with oxygen gas in the water apparatus, Pl. V. Fig. 1.; when it is perfectly full, so that the gas begins to flow out below, the jar, A, is carried to the mercury apparatus, Pl. IV. Fig. 3. We then dry the surface of the mercury, both within and without the jar, by means of blotting-paper, taking care to keep the paper for some time entirely immersed in the mercury before it is introduced under the jar, lest we let in any common air, which sticks very obstinately to the surface of the paper. The body to be submitted to combustion, being first very accurately weighed in nice scales, is placed in a small flat shallow dish, D, of iron or porcelain; this is covered by the larger cup P, which serves the office of a diving bell, and the whole is passed through the mercury into the jar, after which the larger cup is retired. The difficulty of passing the materials of combustion in this manner[Pg 419] through the mercury may be avoided by raising one of the sides of the jar, A, for a moment, and slipping in the little cup, D, with the combustible body as quickly as possible. In this manner of operating, a small quantity of common air gets into the jar, but it is so very inconsiderable as not to injure either the progress or accuracy of the experiment in any sensible degree.

When the cup, D, is introduced under the jar, we suck out a part of the oxygen gas, so as to raise the mercury to EF, as formerly directed, Part I. Chap. V. otherwise, when the combustible body is set on fire, the gas becoming dilated would be in part forced out, and we should no longer be able to make any accurate calculation of the quantities before and after the experiment. A very convenient mode of drawing out the air is by means of an air-pump syringe adapted to the syphon, GHI, by which the mercury may be raised to any degree under twenty-eight inches. Very inflammable bodies, as phosphorus, are set on fire by means of the crooked iron wire, MN, Pl. IV. Fig. 16. made red hot, and passed quickly through the mercury. Such as are less easily set on fire have a small portion of tinder, upon which a minute particle of phosphorus is fixed, laid upon them before using the red hot iron.[Pg 420]

In the first moment of combustion the air, being heated, rarifies, and the mercury descends; but when, as in combustions of phosphorus and iron, no elastic fluid is formed, absorption becomes presently very sensible, and the mercury rises high into the jar. Great attention must be used not to burn too large a quantity of any substance in a given quantity of gas, otherwise, towards the end of the experiment, the cup would approach so near the top of the jar as to endanger breaking it by the great heat produced, and the sudden refrigeration from the cold mercury. For the methods of measuring the volume of the gasses, and for correcting the measures according to the heighth of the barometer and thermometer, &c. see Chap. II. Sect. V. and VI. of this part.

The above process answers very well for burning all the concrete substances, and even for the fixed oils: These last are burnt in lamps under the jar, and are readily set on fire by means of tinder, phosphorus, and hot iron. But it is dangerous for substances susceptible of evaporating in a moderate heat, such as ether, alkohol, and the essential oils; these substances dissolve in considerable quantity in oxygen gas; and, when set on fire, a dangerous and sudden explosion takes place, which carries up the jar to a great height, and dashes it in a thousand pieces. From two such explosions some of the[Pg 421] members of the Academy and myself escaped very narrowly. Besides, though this manner of operating is sufficient for determining pretty accurately the quantity of oxygen gas absorbed, and of carbonic acid produced, as water is likewise formed in all experiments upon vegetable and animal matters which contain an excess of hydrogen, this apparatus can neither collect it nor determine its quantity. The experiment with phosphorus is even incomplete in this way, as it is impossible to demonstrate that the weight of the phosphoric acid produced is equal to the sum of the weights of the phosphorus burnt and oxygen gas absorbed during the process. I have been therefore obliged to vary the instruments according to circumstances, and to employ several of different kinds, which I shall describe in their order, beginning with that used for burning phosphorus.

Take a large balloon, A, Pl. IV. Fig. 4. of cristal or white glass, with an opening, EF, about two inches and a half, or three inches, diameter, to which a cap of brass is accurately fitted with emery, and which has two holes for the passage of the tubes xxx, yyy. Before shutting the balloon with its cover, place within it the stand, BC, supporting the cup of porcelain, D, which contains the phosphorus. Then lute on the cap with fat lute, and allow it to dry for some days, and weigh the whole accurately;[Pg 422]after this exhaust the balloon by means of an air-pump connected with the tube xxx, and fill it with oxygen gas by the tube yyy, from the gazometer, Pl. VIII. Fig. 1. described Chap. II. Sect II. of this part. The phosphorus is then set on fire by means of a burning-glass, and is allowed to burn till the cloud of concrete phosphoric acid stops the combustion, oxygen gas being continually supplied from the gazometer. When the apparatus has cooled, it is weighed and unluted; the tare of the instrument being allowed, the weight is that of the phosphoric acid contained. It is proper, for greater accuracy, to examine the air or gas contained in the balloon after combustion, as it may happen to be somewhat heavier or lighter than common air; and this difference of weight must be taken into account in the calculations upon the results of the experiment.

SECT. III.
Of the Combustion of Charcoal.

The apparatus I have employed for this process consists of a small conical furnace of hammered copper, represented in perspective, Pl. XII. Fig. 9. and internally displayed Fig. 11. It is[Pg 423] divided into the furnace, ABC, where the charcoal is burnt, the grate, $d\ e$, and the ash-hole, F; the tube, GH, in the middle of the dome of the furnace serves to introduce the charcoal, and as a chimney for carrying off the air which has served for combustion. Through the tube, $l\ m\ n$, which communicates with the gazometer, the hydrogen gas, or air, intended for supporting the combustion, is conveyed into the ash-hole, F, whence it is forced, by the application of pressure to the gazometer, to pass through the grate, $d\ e$, and to blow upon the burning charcoal placed immediately above.

Oxygen gas, which forms 28/100 of atmospheric air, is changed into carbonic acid gas during combustion with charcoal, whilst the azotic gas of the air is not altered at all. Hence, after the combustion of charcoal in atmospheric air, a mixture of carbonic acid gas and azotic gas must remain; to allow this mixture to pass off, the tube, *o p*, is adapted to the chimney, GH, by means of a screw at G, and conveys the gas into bottles half filled with solution of caustic potash. The carbonic acid gas is absorbed by the alkali, and the azotic gas is conveyed into a second gazometer, where its quantity is ascertained.

The weight of the furnace, ABC, is first accurately determined, then introduce the tube RS, of known weight, by the chimney, GH,[Pg 424] till its lower end S, rests upon the grate, *d e*, which it occupies entirely; in the next place, fill the furnace with charcoal, and weigh the whole again, to know the exact quantity of charcoal submitted to experiment. The furnace is now put in its place, the tube, *l m n*, is screwed to that which communicates with the gazometer, and the tube, *o p*, to that which communicates with the bottles of alkaline solution. Every thing being in readiness, the stop-cock of the gazometer is opened, a small piece of burning charcoal is thrown into the tube, RS, which is instantly withdrawn, and the tube, *o p*, is screwed to the chimney, GH. The little piece of charcoal falls upon the grate, and in this manner gets below the whole charcoal, and is kept on fire by the stream of air from the gazometer. To be certain that the combustion is begun, and goes on properly, the tube, *q r s*, is fixed to the furnace, having a piece of glass cemented to its upper extremity, *s*, through which we can see if the charcoal be on fire.

I neglected to observe above, that the furnace, and its appendages, are plunged in water in the cistern, TVXY, Fig. 11. Pl. XII. to which ice may be added to moderate the heat, if necessary; though the heat is by no means very considerable, as there is no air but what comes from the gazometer, and no more of the charcoal[Pg 425] burns at one time than what is immediately over the grate.

As one piece of charcoal is consumed another falls down into its place, in consequence of the declivity of the sides of the furnace; this gets into the stream of air from the grate, *d e*, and is burnt; and so on, successively, till the whole charcoal is consumed. The air which has served the purpose of the combustion passes through the mass of charcoal, and is forced by the pressure of the gazometer to escape through the tube, *o p*, and to pass through the bottles of alkaline solution.

This experiment furnishes all the necessary data for a complete analysis of atmospheric air and of charcoal. We know the weight of charcoal consumed; the gazometer gives us the measure of the air employed; the quantity and quality of gas remaining after combustion may be determined, as it is received, either in another gazometer, or in jars, in a pneumato-chemical apparatus; the weight of ashes remaining in the ash-hole is readily ascertained; and, finally, the additional weight acquired by the bottles of alkaline solution gives the exact quantity of carbonic acid formed during the process. By this experiment we may likewise determine, with sufficient accuracy, the proportions in which charcoal and oxygen enter into the composition of carbonic acid.[Pg 426]

In a future memoir I shall give an account to the Academy of a series of experiments I have undertaken, with this instrument, upon all the vegetable and animal charcoals. By some very slight alterations, this machine may be made to answer for observing the principal phenomena of respiration.

SECT. IV.
Of the Combustion of Oils.

Oils are more compound in their nature than charcoal, being formed by the combination of at least two elements, charcoal and hydrogen; of course, after their combustion in common air, water, carbonic acid gas, and azotic gas, remain. Hence the apparatus employed for their combustion requires to be adapted for collecting these three products, and is consequently more complicated than the charcoal furnace.

The apparatus I employ for this purpose is composed of a large jar or pitcher A, Pl. XII. Fig. 4. surrounded at its upper edge by a rim of iron properly cemented at DE, and receding from the jar at BC, so as to leave a furrow or gutter *xx*, between it and the outside of the jar,[Pg 427] somewhat more than two inches deep. The cover or lid of the jar, Fig. 5. is likewise surrounded by an iron rim *f g*, which adjusts into the gutter *xx*, Fig. 4. which being filled with mercury, has the effect of closing the jar hermetically in an instant, without using any lute; and, as the gutter will hold about two inches of mercury, the air in the jar may be made to sustain the pressure of more than two feet of water, without danger of its escaping.

The lid has four holes, T *h i k*, for the passage of an equal number of tubes. The opening T is furnished with a leather box, through which passes the rod, Fig. 3. intended for raising and lowering the wick of the lamp, as will be afterwards directed. The three other holes are intended for the passage of three several tubes, one of which conveys the oil to the lamp, a second conveys air for keeping up the combustion, and the third carries off the air, after it has served for combustion. The lamp in which the oil is burnt is represented Fig. 2; *a* is the reservoir of oil, having a funnel by which it is filled; *b c d e f g h* is a syphon which conveys the oil to the lamp 11; 7, 8, 9, 10, is the tube which conveys the air for combustion from the gazometer to the same lamp. The tube *b c* is formed externally, at its lower end *b*, into a male screw, which turns in a female screw in the lid of the reservoir of oil *a*; so that, by turning[Pg 428] the reservoir one way or the other, it is made to rise or fall, by which the oil is kept at the necessary level.

When the syphon is to be filled, and the communication formed between the reservoir of oil and the lamp, the stop-cock *c* is shut, and that at *e* opened, oil is poured in by the opening *f* at the top of the syphon, till it rises within three or four lines of the upper edge of the lamp, the stop-cock *k* is then shut, and that at *c* opened; the oil is then poured in at *f*, till the branch *b c d* of the syphon is filled, and then the stop-cock *e* is closed. The two branches of the syphon being now completely filled, a communication is fully established between the reservoir and the lamp.

In Pl. XII. Fig. 1. all the parts of the lamp 11, Fig. 2. are represented magnified, to show them distinctly. The tube *i k* carries the oil from the reservoir to the cavity *a a a a*, which contains the wick; the tube 9, 10, brings the air from the gazometer for keeping up the combustion; this air spreads through the cavity *d d d d*, and, by means of the passages *c c c c* and *b b b b*, is distributed on each side of the wick, after the principles of the lamps constructed by Argand, Quinquet, and Lange.

To render the whole of this complicated apparatus more easily understood, and that its description may make all others of the same kind[Pg 429] more readily followed, it is represented, completely connected together for use, in Pl. XI. The gazometer P furnishes air for the combustion by the tube and stop-cock 1, 2; the tube 2, 3, communicates with a second gazometer, which is filled whilst the first one is emptying during the process, that there may be no interruption to the combustion; 4, 5, is a tube of glass filled with deliquescent salts, for drying the air as much as possible in its passage; and the weight of this tube and its contained salts, at the beginning of the experiment, being known, it is easy to determine the quantity of water absorbed by them from the air. From this deliquescent tube the air is conducted through the pipe 5, 6, 7, 8, 9, 10, to the lamp 11, where it spreads on both sides of the wick, as before described, and feeds the flame. One part of this air, which serves to keep up the combustion of the oil, forms carbonic acid gas and water, by oxygenating its elements. Part of this water condenses upon the sides of the pitcher A, and another part is held in solution in the air by means of caloric furnished by the combustion. This air is forced by the compression of the gazometer to pass through the tube 12, 13, 14, 15, into the bottle 16, and the worm 17, 18, where the water is fully condensed from the refrigeration of the air; and, if any water still remains[Pg 430] in solution, it is absorbed by deliquescent salts contained in the tube 19, 20.

All these precautions are solely intended for collecting and determining the quantity of water formed during the experiment; the carbonic acid and azotic gas remains to be ascertained. The former is absorbed by caustic alkaline solution in the bottles 22 and 25. I have only represented two of these in the figure, but nine at least are requisite; and the last of the series may be half filled with lime-water, which is the most certain reagent for indicating the presence of carbonic acid; if the lime-water is not rendered turbid, we may be certain that no sensible quantity of that acid remains in the air.

The rest of the air which has served for combustion, and which chiefly consists of azotic gas, though still mixed with a considerable portion of oxygen gas, which has escaped unchanged from the combustion, is carried through a third tube 28, 29, of deliquescent salts, to deprive it of any moisture it may have acquired in the bottles of alkaline solution and lime-water, and from thence by the tube 29, 30, into a gazometer, where its quantity is ascertained. Small essays are then taken from it, which are exposed to a solution of sulphuret of potash, to ascertain the proportions of oxygen and azotic gas it contains.[Pg 431]

In the combustion of oils the wick becomes charred at last, and obstructs the rise of the oil; besides, if we raise the wick above a certain height, more oil rises through its capillary tubes than the stream of air is capable of consuming, and smoke is produced. Hence it is necessary to be able to lengthen or shorten the wick without opening the apparatus; this is accomplished by means of the rod 31, 32, 33, 34, which passes through a leather-box, and is connected with the support of the wick; and that the motion of this rod, and consequently of the wick, may be regulated with the utmost smoothness and facility; it is moved at pleasure by a pinnion which plays in a toothed rack. The rod, with its appendages, are represented Pl. XII. Fig. 3. It appeared to me, that the combustion would be assisted by surrounding the flame of the lamp with a small glass jar open at both ends, as represented in its place in Pl. XI.

I shall not enter into a more detailed description of the construction of this apparatus, which is still capable of being altered and modified in many respects, but shall only add, that when it is to be used in experiment, the lamp and reservoir with the contained oil must be accurately weighed, after which it is placed as before directed, and lighted; having then formed the connection between the air in the gazometer and the lamp, the external jar A, Pl. XI. is fixed[Pg 432] over all, and secured by means of the board BC and two rods of iron which connect this board with the lid, and are screwed to it. A small quantity of oil is burnt while the jar is adjusting to the lid, and the product of that combustion is lost; there is likewise a small portion of air from the gazometer lost at the same time. Both of these are of very inconsiderable consequence in extensive experiments, and they are even capable of being valued in our calculation of the results.

In a particular memoir, I shall give an account to the Academy of the difficulties inseparable from this kind of experiments: These are so insurmountable and troublesome, that I have not hitherto been able to obtain any rigorous determination of the quantities of the products. I have sufficient proof, however, that the fixed oils are entirely resolved during combustion into water and carbonic acid gas, and consequently that they are composed of hydrogen and charcoal; but I have no certain knowledge respecting the proportions of these ingredients.[Pg 433]

SECT. V.
Of the Combustion of Alkohol.

The combustion of alkohol may be very readily performed in the apparatus already described for the combustion of charcoal and phosphorus. A lamp filled with alkohol is placed under the jar A, Pl. IV. Fig. 3. a small morsel of phosphorus is placed upon the wick of the lamp, which is set on fire by means of the hot iron, as before directed. This process is, however, liable to considerable inconveniency; it is dangerous to make use of oxygen gas at the beginning of the experiment for fear of deflagration, which is even liable to happen when common air is employed. An instance of this had very near proved fatal to myself, in presence of some members of the Academy. Instead of preparing the experiment, as usual, at the time it was to be performed, I had disposed every thing in order the evening before; the atmospheric air of the jar had thereby sufficient time to dissolve a good deal of the alkohol; and this evaporation had even been considerably promoted by the height of the column of mercury, which I had raised to EF, Pl. IV. Fig. 3. The moment I attempted[Pg 434] to set the little morsel of phosphorus on fire by means of the red hot iron, a violent explosion took place, which threw the jar with great violence against the floor of the laboratory, and dashed it in a thousand pieces.

Hence we can only operate upon very small quantities, such as ten or twelve grains of alkohol, in this manner; and the errors which may be committed in experiments upon such small quantities prevents our placing any confidence in their results. I endeavoured to prolong the combustion, in the experiments contained in the Memoirs of the Academy for 1784. p. 593. by lighting the alkohol first in common air, and furnishing oxygen gas afterwards to the jar, in proportion as it consumed; but the carbonic acid gas produced by the process became a great hinderance to the combustion, the more so that alkohol is but difficultly combustible, especially in worse than common air; so that even in this way very small quantities only could be burnt.

Perhaps this combustion might succeed better in the oil apparatus, Pl. XI.; but I have not hitherto ventured to try it. The jar A in which the combustion is performed is near 1400 cubical inches in dimension; and, were an explosion to take place in such a vessel, its consequences would be very terrible, and very difficult to guard against. I have not, however, despaired of making the attempt.[Pg 435]

From all these difficulties, I have been hitherto obliged to confine myself to experiments upon very small quantities of alkohol, or at least to combustions made in open vessels, such as that represented in Pl. IX. Fig. 5. which will be described in Section VII. of this chapter. If I am ever able to remove these difficulties, I shall resume this investigation.

SECT. VI.
Of the Combustion of Ether.

Tho' the combustion of ether in close vessels does not present the same difficulties as that of alkohol, yet it involves some of a different kind, not more easily overcome, and which still prevent the progress of my experiments. I endeavoured to profit by the property which ether possesses of dissolving in atmospheric air, and rendering it inflammable without explosion. For this purpose, I constructed the reservoir of ether *a b c d*, Plate XII. Fig. 8. to which air is brought from the gazometer by the tube 1, 2, 3, 4. This air spreads, in the first place, in the double lid *ac* of the reservoir, from which it passes through seven tubes *ef, gh, ik*, &c. which descend to the bottom of the ether, and it is[Pg 436] forced by the pressure of the gazometer to boil up through the ether in the reservoir. We may replace the ether in this first reservoir, in proportion as it is dissolved and carried off by the air, by means of the supplementary reservoir E, connected by a brass tube fifteen or eighteen inches long, and shut by a stop-cock. This length of the connecting tube is to enable the descending ether to overcome the resistance occasioned by the pressure of the air from the gazometer.

The air, thus loaded with vapours of ether, is conducted by the tube 5, 6, 7, 8, 9, to the jar A, into which it is allowed to escape through a capillary opening, at the extremity of which it is set on fire. The air, when it has served the purpose of combustion, passes through the bottle 16, Pl. XI. the worm 17, 18, and the deliquescent tube 19, 20, after which it passes through the alkaline bottles; in these its carbonic acid gas is absorbed, the water formed during the experiment having been previously deposited in the former parts of the apparatus.

When I caused construct this apparatus, I supposed that the combination of atmospheric air and ether formed in the reservoir *a b c d*, Pl. XII. Fig. 8. was in proper proportion for supporting combustion; but in this I was mistaken; for there is a very considerable quantity of excess of ether; so that an additional quantity of atmospheric[Pg 437]air is necessary to enable it to burn fully. Hence a lamp constructed upon these principles will burn in common air, which furnishes the quantity of oxygen necessary for combustion, but will not burn in close vessels in which the air is not renewed. From this circumstance, my ether lamp went out soon after being lighted and shut up in the jar A, Pl. XII. Fig. 8. To remedy this defect, I endeavoured to bring atmospheric air to the lamp by the lateral tube 10, 11, 12, 13, 14, 15, which I distributed circularly round the flame; but the flame is so exceedingly rare, that it is blown out by the gentlest possible stream of air, so that I have not hitherto succeeded in burning ether. I do not, however, despair of being able to accomplish it by means of some changes I am about to have made upon this apparatus.

SECT. VII.
Of the Combustion of Hydrogen Gas, and the Formation of Water.

In the formation of water, two substances, hydrogen and oxygen, which are both in the aëriform state before combustion, are transformed into liquid or water by the operation.[Pg 438] This experiment would be very easy, and would require very simple instruments, if it were possible to procure the two gasses perfectly pure, so that they might burn without any residuum. We might, in that case, operate in very small vessels, and, by continually furnishing the two gasses in proper proportions, might continue the combustion indefinitely. But, hitherto, chemists have only employed oxygen gas, mixed with azotic gas;

from which circumstance, they have only been able to keep up the combustion of hydrogen gas for a very limited time in close vessels, because, as the residuum of azotic gas is continually increasing, the air becomes at last so much contaminated, that the flame weakens and goes out. This inconvenience is so much the greater in proportion as the oxygen gas employed is less pure. From this circumstance, we must either be satisfied with operating upon small quantities, or must exhaust the vessels at intervals, to get rid of the residuum of azotic gas; but, in this case, a portion of the water formed during the experiment is evaporated by the exhaustion; and the resulting error is the more dangerous to the accuracy of the process, that we have no certain means of valuing it.

These considerations make me desirous to repeat the principal experiments of pneumatic chemistry with oxygen gas entirely free from[Pg 439] any admixture of azotic gas; and this may be procured from oxygenated muriat of potash. The oxygen gas extracted from this salt does not appear to contain azote, unless accidentally, so that, by proper precautions, it may be obtained perfectly pure. In the mean time, the apparatus employed by Mr Meusnier and me for the combustion of hydrogen gas, which is described in the experiment for recomposition of water, Part I. Chap. VIII. and need not be here repeated, will answer the purpose; when pure gasses are procured, this apparatus will require no alterations, except that the capacity of the vessels may then be diminished. See Pl. IV. Fig. 5.

The combustion, when once begun, continues for a considerable time, but weakens gradually, in proportion as the quantity of azotic gas remaining from the combustion increases, till at last the azotic gas is in such over proportion that the combustion can no longer be supported, and the flame goes out. This spontaneous extinction must be prevented, because, as the hydrogen gas is pressed upon in its reservoir, by an inch and a half of water, whilst the oxygen gas suffers a pressure only of three lines, a mixture of the two would take place in the balloon, which would at last be forced by the superior pressure into the reservoir of oxygen gas. Wherefore the combustion must be stopped,[Pg 440] by shutting the stop-cock of the tube dDd whenever the flame grows very feeble; for which purpose it must be attentively watched.

There is another apparatus for combustion, which, though we cannot with it perform experiments with the same scrupulous exactness as with the preceding instruments, gives very striking results that are extremely proper to be shewn in courses of philosophical chemistry. It consists of a worm EF, Pl. IX. Fig. 5. contained in a metallic cooller ABCD. To the upper part of this worm E, the chimney GH is fixed, which is composed of two tubes, the inner of which is a continuation of the worm, and the outer one is a case of tin-plate, which surrounds it at about an inch distance, and the interval is filled up with sand. At the inferior extremity K of the inner tube, a glass tube is fixed, to which we adopt the Argand lamp LM for burning alkohol, &c.

Things being thus disposed, and the lamp being filled with a determinate quantity of alkohol, it is set on fire; the water which is formed during the combustion rises in the chimney KE, and being condensed in the worm, runs out at its extremity F into the bottle P. The double tube of the chimney, filled with sand in the interstice, is to prevent the tube from cooling in its upper part, and condensing the water; otherwise,[Pg 441] it would fall back in the tube, and we should not be able to ascertain its quantity, and besides it might fall in drops upon the wick, and extinguish the flame. The intention of this construction, is to keep the chimney always hot, and the worm always cool, that the water may be preserved in the state of vapour whilst rising, and may be condensed immediately upon getting into the descending part of the apparatus. By this instrument, which was contrived by Mr Meusnier, and which is described by me in the Memoirs of the Academy for 1784, p. 593. we may, with attention to keep the worm always cold, collect nearly seventeen ounces of water from the combustion of sixteen ounces of alkohol.

SECT. VIII.
Of the Oxydation of Metals.

The term *oxydation* or *calcination* is chiefly used to signify the process by which metals exposed to a certain degree of heat are converted into oxyds, by absorbing oxygen from the air. This combination takes place in consequence of oxygen possessing a greater affinity to metals, at a certain temperature, than to caloric, which[Pg 442]becomes disengaged in its free state; but, as this disengagement, when made in common air, is slow and progressive, it is scarcely evident to the senses. It is quite otherwise, however, when oxydation takes place in oxygen gas; for, being produced with much greater rapidity, it is generally accompanied with heat and light, so as evidently to show that metallic substances are real combustible bodies.

All the metals have not the same degree of affinity to oxygen. Gold, silver, and platina, for instance, are incapable of taking it away from its combination with caloric, even in the greatest known heat; whereas the other metals absorb it in a larger or smaller quantity, until the affinities of the metal to oxygen, and of the latter to caloric, are in exact equilibrium. Indeed, this state of equilibrium of affinities may be assumed as a general law of nature in all combinations.

In all operations of this nature, the oxydation of metals is accelerated by giving free access to the air; it is sometimes much assisted by joining the action of a bellows, which directs a stream of air over the surface of the metal. This process becomes greatly more rapid if a stream of oxygen gas be used, which is readily done by means of the gazometer formerly described. The metal, in this case, throws out a brilliant flame, and the oxydation is very quickly[Pg 443] accomplished; but this method can only be used in very confined experiments, on account of the expence of procuring oxygen gas. In the essay of ores, and in all the common operations of the laboratory, the calcination or oxydation of metals is usually performed in a dish of baked clay, Pl. IV. Fig. 6. commonly called a *roasting test*, placed in a strong furnace. The substances to be oxydated are frequently stirred, on purpose to present fresh surfaces to the air.

Whenever this operation is performed upon a metal which is not volatile, and from which nothing flies off into the surrounding air during the process, the metal acquires additional weight; but the cause of this increased weight during oxydation could never have been discovered by means of experiments performed in free air; and it is only since these operations have been performed in close vessels, and in determinate quantities of air, that any just conjectures have been formed concerning the cause of this phenomenon. The first method for this purpose is due to Dr Priestley, who exposes the metal to be calcined in a porcelain cup N, Pl. IV. Fig. 11. placed upon the stand IK, under a jar A, in the bason BCDE, full of water; the water is made to rise up to GH, by sucking out the air with a syphon, and the focus of a burning glass is made to fall upon the metal. In a few minutes the oxydation takes place,[Pg 444] a part of the oxygen contained in the air combines with the metal, and a proportional diminution of the volume of air is produced; what remains is nothing more than azotic gas, still however mixed with a small quantity of oxygen gas. I have given an account of a series of experiments made with this apparatus in my Physical and Chemical Essays, first published in 1773. Mercury may be used instead of water in this experiment, whereby the results are rendered still more conclusive.

Another process for this purpose was invented by Mr Boyle, and of which I gave an account in the Memoirs of the Academy for 1774, p. 351. The metal is introduced into a retort, Pl. III. Fig. 20. the beak of which is hermetically sealed; the metal is then oxydated by means of heat applied with great precaution. The weight of the vessel, and its contained substances, is not at all changed by this process, until the extremity of the neck of the retort is broken; but, when that is done, the external air rushes in with a hissing noise. This operation is attended with danger, unless a part of the air is driven out of the retort, by means of heat, before it is hermetically sealed, as otherwise the retort would be apt to burst by the dilation of the air when placed in the furnace. The quantity of air driven out may be received under a jar in the pneumato-chemical apparatus,[Pg 445] by which its quantity, and that of the air remaining in the retort, is ascertained. I have not multiplied my experiments upon oxydation of metals so much as I could have wished; neither have I obtained satisfactory results with any metal except tin. It is much to be wished that some person would undertake a series of experiments upon oxydation of metals in the several gasses; the subject is important, and would fully repay any trouble which this kind of experiment might occasion.

As all the oxyds of mercury are capable of revivifying without addition, and restore the oxygen gas they had before absorbed, this seemed to be the most proper metal for becoming the subject of conclusive experiments upon oxydation. I formerly endeavoured to accomplish the oxydation of mercury in close vessels, by filling a retort, containing a small quantity of mercury, with oxygen gas, and adapting a bladder half full of the same gas to its beak; See Pl. IV. Fig. 12. Afterwards, by heating the mercury in the retort for a very long time, I succeeded in oxydating a very small portion, so as to form a little red oxyd floating upon the surface of the running mercury; but the quantity was so small, that the smallest error committed in the determination of the quantities of oxygen gas before and after the operation must have thrown very great uncertainty upon the[Pg 446] results of the experiment. I was, besides, dissatisfied with this process, and not without cause, lest any air might have escaped through the pores of the bladder, more especially as it becomes shrivelled by the heat of the furnace, unless covered over with cloths kept constantly wet.

This experiment is performed with more certainty in the apparatus described in the Memoirs of the Academy for 1775, p. 580. This consists of a retort, A, Pl. IV. Fig. 2. having a crooked glass tube BCDE of ten or twelve lines internal diameter, melted on to its beak, and which is engaged under the bell glass FG, standing with its mouth downwards, in a bason filled with water or mercury. The retort is placed upon the bars of the furnace MMNN, Pl. IV. Fig. 2. or in a sand bath, and by means of this apparatus we may, in the course of several days, oxydate a small quantity of mercury in common air; the red oxyd floats upon the surface, from which it may be collected and revivified, so as to compare the quantity of oxygen gas obtained in revivification with the absorption which took place during oxydation. This kind of experiment can only be performed upon a small scale, so that no very certain conclusions can be drawn from them[61].

[Pg 447]
The combustion of iron in oxygen gas being a true oxydation of that metal, ought to be mentioned in this place. The apparatus employed by Mr Ingenhousz for this operation is represented in Pl. IV. Fig. 17.; but, having already described it sufficiently in Chap. III. I shall refer the reader to what is said of it in that place. Iron may likewise be oxydated by combustion in vessels filled with oxygen gas, in the way already directed for phosphorus and charcoal. This apparatus is represented Pl. IV. Fig. 3. and described in the fifth chapter of the first part of this work. We learn from Mr Ingenhousz, that all the metals, except gold, silver, and mercury, may be burnt or oxydated in the same manner, by reducing them into very fine wire, or very thin plates cut into narrow slips; these are twisted round with iron-wire, which communicates the property of burning to the other metals.

Mercury is even difficultly oxydated in free air. In chemical laboratories, this process is usually carried on in a matrass A, Pl. IV. Fig. having a very flat body, and a very long neck BC, which vessel is commonly called *Boyle's bell*. A quantity of mercury is introduced sufficient to cover the bottom, and it is placed in a sand-bath, which keeps up a constant heat approaching to that of boiling mercury. By continuing this operation with five or six similar matrasses during several months, and renewing[Pg 448] the mercury from time to time, a few ounces of red oxyd are at last obtained. The great slowness and inconvenience of this apparatus arises from the air not being sufficiently renewed; but if, on the other hand, too free a circulation were given to the external air, it would carry off the mercury in solution in the state of vapour, so that in a few days none would remain in the vessel.

As, of all the experiments upon the oxydation of metals, those with mercury are the most conclusive, it were much to be wished that a simple apparatus could be contrived by which this oxydation and its results might be demonstrated in public courses of chemistry. This might, in my opinion, be accomplished by methods similar to those I have already described for the combustion of charcoal and the oils; but, from other pursuits, I have not been able hitherto to resume this kind of experiment.

The oxyd of mercury revives without addition, by being heated to a slightly red heat. In this degree of temperature, oxygen has greater affinity to caloric than to mercury, and forms oxygen gas. This is always mixed with a small portion of azotic gas, which indicates that the mercury absorbs a small portion of this latter gas during oxydation. It almost always contains a little carbonic acid gas, which must undoubtedly be attributed to the foulnesses of the[Pg 449] oxyd; these are charred by the heat, and convert a part of the oxygen gas into carbonic acid.

If chemists were reduced to the necessity of procuring all the oxygen gas employed in their experiments from mercury oxydated by heat without addition, or, as it is called,*calcined* or *precipitated per se*, the excessive dearness of that preparation would render experiments, even upon a moderate scale, quite impracticable. But mercury may likewise be oxydated by means of nitric acid; and in this way we procure a red oxyd, even more pure than that produced by calcination. I have sometimes prepared this oxyd by dissolving mercury in nitric acid, evaporating to dryness, and calcining the salt, either in a retort, or in capsules formed of pieces of broken matrasses and retorts, in the manner formerly described; but I have never succeeded in making it equally beautiful with what is sold by the druggists, and which is, I believe, brought from Holland. In choosing this, we ought to prefer what is in solid lumps composed of soft adhering scales, as when in powder it is sometimes adulterated with red oxyd of lead.

To obtain oxygen gas from the red oxyd of mercury, I usually employ a porcelain retort, having a long glass tube adapted to its beak, which is engaged under jars in the water pneumato-chemical[Pg 450] apparatus, and I place a bottle in the water, at the end of the tube, for receiving the mercury, in proportion as it revives and distils over. As the oxygen gas never appears till the retort becomes red, it seems to prove the principle established by Mr Berthollet, that an obscure heat can never form oxygen gas, and that light is one of its constituent elements. We must reject the first portion of gas which comes over, as being mixed with common air, from what was contained in the retort at the beginning of the experiment; but, even with this precaution, the oxygen gas procured is usually contaminated with a tenth part of azotic gas, and with a very small portion of carbonic acid gas. This latter is readily got rid of, by making the gas pass through a solution of caustic alkali; but we know of no method for separating the azotic gas; its proportions may however be ascertained, by leaving a known quantity of the oxygen gas contaminated with it for a fortnight, in contact with sulphuret of soda or potash, which absorbs the oxygen gas so as to convert the sulphur into sulphuric acid, and leaves the azotic gas remaining pure.

We may likewise procure oxygen gas from black oxyd of manganese or nitrat of potash, by exposing them to a red heat in the apparatus already described for operating upon red[Pg 451] oxyd of mercury; only, as it requires such a heat as is at least capable of softening glass, we must employ retorts of stone or of porcelain. But the purest and best oxygen gas is what is disengaged from oxygenated muriat of potash by simple heat. This operation is performed in a glass retort, and the gas obtained is perfectly pure, provided that the first portions, which are mixed with the common air of the vessels, be rejected.

FOOTNOTES:
[61]See an account of this experiment, Part. I. Chap. iii.—A.

[Pg 452]

CHAP. IX.
Of Deflagration.

I have already shown, Part I. Chap. IX. that oxygen does not always part with the whole of the caloric it contained in the state of gas when it enters into combination with other bodies. It carries almost the whole of its caloric alongst with it in entering into the combinations which form nitric acid and oxygenated muriatic acid; so that in nitrats, and more especially in oxygenated muriats, the oxygen is, in a certain degree, in the state of oxygen gas, condensed, and reduced to the smallest volume it is capable of occupying.

In these combinations, the caloric exerts a constant action upon the oxygen to bring it back to the state of gas; hence the oxygen adheres but very slightly, and the smallest additional force is capable of setting it free; and, when such force is applied, it often recovers the state of gas instantaneously. This rapid passage from the solid to the aëriform state is called detonation, or fulmination, because it is usually accompanied with noise and explosion. Deflagrations are commonly produced by means of combinations of charcoal either with nitre or[Pg 453] oxygenated muriat of potash; sometimes, to assist the inflammation, sulphur is added; and, upon the just proportion of these ingredients, and the proper manipulation of the mixture, depends the art of making gun-powder.

As oxygen is changed, by deflagration with charcoal, into carbonic acid, instead of oxygen gas, carbonic acid gas is disengaged, at least when the mixture has been made in just proportions. In deflagration with nitre, azotic gas is likewise disengaged, because azote is one of the constituent elements of nitric acid.

The sudden and instantaneous disengagement and expansion of these gasses is not, however, sufficient for explaining all the phenomena of deflagration; because, if this were the sole operating power, gun powder would always be so much the stronger in proportion as the quantity of gas disengaged in a given time was the more considerable, which does not always accord with experiment. I have tried some kinds which produced almost double the effect of ordinary gun powder, although they gave out a sixth part less of gas during deflagration. It would appear that the quantity of caloric disengaged at the moment of detonation contributes considerably to the expansive effects produced; for, although caloric penetrates freely through the pores of every body in nature, it can only do so progressively, and in a given time; hence,[Pg 454]when the quantity disengaged at once is too large to get through the pores of the surrounding bodies, it must necessarily act in the same way with ordinary elastic fluids, and overturn every thing that opposes its passage. This must, at least in part, take place when gun-powder is set on fire in a cannon; as, although the metal is permeable to caloric, the quantity disengaged at once is too large to find its way through the pores of the metal, it must therefore make an effort to escape on every side; and, as the resistance all around, excepting towards the muzzle, is too great to be overcome, this effort is employed for expelling the bullet.

The caloric produces a second effect, by means of the repulsive force exerted between its particles; it causes the gasses, disengaged at the moment of deflagration, to expand with a degree of force proportioned to the temperature produced.

It is very probable that water is decomposed during the deflagration of gun-powder, and that part of the oxygen furnished to the nascent carbonic acid gas is produced from it. If so, a considerable quantity of hydrogen gas must be disengaged in the instant of deflagration, which expands, and contributes to the force of the explosion. It may readily be conceived how greatly this circumstance must increase the effect of powder, if we consider that a pint of hydrogen[Pg 455] gas weighs only one grain and two thirds; hence a very small quantity in weight must occupy a very large space, and it must exert a prodigious expansive force in passing from the liquid to the aëriform state of existence.

In the last place, as a portion of undecomposed water is reduced to vapour during the deflagration of gun-powder, and as water, in the state of gas, occupies seventeen or eighteen hundred times more space than in its liquid state, this circumstance must likewise contribute largely to the explosive force of the powder.

I have already made a considerable series of experiments upon the nature of the elastic fluids disengaged during the deflagration of nitre with charcoal and sulphur; and have made some, likewise, with the oxygenated muriat of potash. This method of investigation leads to tollerably accurate conclusions with respect to the constituent elements of these salts. Some of the principal results of these experiments, and of the consequences drawn from them respecting the analysis of nitric acid, are reported in the collection of memoirs presented to the Academy by foreign philosophers, vol. xi. p. 625. Since then I have procured more convenient instruments, and I intend to repeat these experiments upon a larger scale, by which I shall procure more accurate precision in their results; the following, however, is the process I have hitherto[Pg 456]employed. I would very earnestly advise such as intend to repeat some of these experiments, to be very much upon their guard in operating upon any mixture which contains nitre, charcoal, and sulphur, and more especially with those in which oxygenated muriat of potash is mixed with these two materials.

I make use of pistol barrels, about six inches long, and of five or six lines diameter, having the touch-hole spiked up with an iron nail strongly driven in, and broken in the hole, and a little tin-smith's solder run in to prevent any possible issue for the air. These are charged with a mixture of known quantities of nitre and charcoal, or any other mixture capable of deflagration, reduced to an impalpable powder, and formed into a paste with a moderate quantity of water. Every portion of the materials introduced must be rammed down with a rammer nearly of the same caliber with the barrel, four or five lines at the muzzle must be left empty, and about two inches of quick match are added at the end of the charge. The only difficulty in this experiment, especially when sulphur is contained in the mixture, is to discover the proper degree of moistening; for, if the paste be too much wetted, it will not take fire, and if too dry, the deflagration is apt to become too rapid, and even dangerous.[Pg 457]

When the experiment is not intended to be rigorously exact, we set fire to the match, and, when it is just about to communicate with the charge, we plunge the pistol below a large bell-glass full of water, in the pneumato chemical apparatus. The deflagration begins, and continues in the water, and gas is disengaged with less or more rapidity, in proportion as the mixture is more or less dry. So long as the deflagration continues, the muzzle of the pistol must be kept somewhat inclined downwards, to prevent the water from getting into its barrel. In this manner I have sometimes collected the gas produced from the deflagration of an ounce and half, or two ounces, of nitre.

In this manner of operating it is impossible to determine the quantity of carbonic acid gas disengaged, because a part of it is absorbed by the water while passing through it; but, when the carbonic acid is absorbed, the azotic gas remains; and, if it be agitated for a few minutes in caustic alkaline solution, we obtain it pure, and can easily determine its volume and weight. We may even, in this way, acquire a tollerably exact knowledge of the quantity of carbonic acid by repeating the experiment a great many times, and varying the proportions of charcoal, till we find the exact quantity requisite to deflagrate the whole nitre employed. Hence, by means of the weight of charcoal employed, we[Pg 458] determine the weight of oxygen necessary for saturation, and deduce the quantity of oxygen contained in a given weight of nitre.

I have used another process, by which the results of this experiment are considerably more accurate, which consists in receiving the disengaged gasses in bell-glasses filled with mercury. The mercurial apparatus I employ is large enough to contain jars of from twelve to fifteen pints in capacity, which are not very readily managed when full of mercury, and even require to be filled by a particular method. When the jar is placed in the cistern of mercury, a glass syphon is introduced, connected with a small air-pump, by means of which the air is exhausted, and the mercury rises so as to fill the jar. After this, the gas of the deflagration is made to pass into the jar in the same manner as directed when water is employed.

I must again repeat, that this species of experiment requires to be performed with the greatest possible precautions. I have sometimes seen, when the disengagement of gas proceeded with too great rapidity, jars filled with more than an hundred and fifty pounds of mercury driven off by the force of the explosion, and broken to pieces, while the mercury was scattered about in great quantities.

When the experiment has succeeded, and the gas is collected under the jar, its quantity in[Pg 459] general, and the nature and quantities of the several species of gasses of which the mixture is composed, are accurately ascertained by the methods already pointed out in the second chapter of this part of my work. I have been prevented from putting the last hand to the experiments I had begun upon deflagration, from their connection with the objects I am at present engaged in; and I am in hopes they will throw considerable light upon the operations belonging to the manufacture of gun-powder.

[Pg 460]

CHAP. X.
Of the Instruments necessary for Operating upon Bodies in very high Temperatures.
SECT. I.
Of Fusion.

We have already seen, that, by aqueous solution, in which the particles of bodies are separated from each other, neither the solvent nor the body held in solution are at all decomposed; so that, whenever the cause of separation ceases, the particles reunite, and the saline substance recovers precisely the same appearance and properties it possessed before solution. Real solutions are produced by fire, or by introducing and accumulating a great quantity of caloric between the particles of bodies; and this species of solution in caloric is usually called *fusion*.

This operation is commonly performed in vessels called crucibles, which must necessarily[Pg 461] be less fusible than the bodies they are intended to contain. Hence, in all ages, chemists have been extremely solicitous to procure crucibles of very refractory materials, or such as are capable of resisting a very high degree of heat. The best are made of very pure clay or of porcelain earth; whereas such as are made of clay mixed with calcareous or silicious earth are very fusible. All the crucibles made in the neighbourhood of Paris are of this kind, and consequently unfit for most chemical experiments. The Hessian crucibles are tolerably good; but the best are made of Limoges earth, which seems absolutely infusible. We have, in France, a great many clays very fit for making crucibles; such, for instance, is the kind used for making melting pots at the glass-manufactory of St Gobin.

Crucibles are made of various forms, according to the operations they are intended to perform. Several of the most common kinds are represented Pl. VII. Fig. 7. 8. 9. and 10. the one represented at Fig. 9. is almost shut at its mouth.

Though fusion may often take place without changing the nature of the fused body, this operation is frequently employed as a chemical means of decomposing and recompounding bodies. In this way all the metals are extracted from their ores; and, by this process, they are revivified,[Pg 462] moulded, and alloyed with each other. By this process sand and alkali are combined to form glass, and by it likewise pastes, or coloured stones, enamels, &c. are formed.

The action of violent fire was much more frequently employed by the ancient chemists than it is in modern experiments. Since greater precision has been employed in philosophical researches, the *humid* has been preferred to the *dry* method of process, and fusion is seldom had recourse to until all the other means of analysis have failed.

SECT. II.
Of Furnaces.

These are instruments of most universal use in chemistry; and, as the success of a great number of experiments depends upon their being well or ill constructed, it is of great importance that a laboratory be well provided in this respect. A furnace is a kind of hollow cylindrical tower, sometimes widened above, Pl. XIII. Fig. 1. ABCD, which must have at least two lateral openings; one in its upper part F, which is the door of the fire-place, and one below, G, leading to the ash-hole. Between these the furnace[Pg 463] is divided by a horizontal grate, intended for supporting the fewel, the situation of which is marked in the figure by the line HI. Though this be the least complicated of all the chemical furnaces, yet it is applicable to a great number of purposes. By it lead, tin, bismuth, and, in general, every substance which does not require a very strong fire, may be melted in crucibles; it will serve for metallic oxydations, for evaporatory vessels, and for sand-baths, as in Pl. III. Fig. 1. and 2. To render it proper for these purposes, several notches, *m m m m*, Pl. XIII. Fig. 1. are made in its upper edge, as otherwise any pan which might be placed over the fire would stop the passage of the air, and prevent the fewel from burning. This furnace can only produce a moderate degree of heat, because the quantity of charcoal it is capable of consuming is limited by the quantity of air which is allowed to pass through the opening G of the ash-hole. Its power might be considerably augmented by enlarging this opening, but then the great stream of air which is convenient for some operations might be hurtful in others; wherefore we must have furnaces of different forms, constructed for different purposes, in our laboratories: There ought especially to be several of the kind now described of different sizes.

The reverberatory furnace, Pl. XIII. Fig. 2. is perhaps more necessary. This, like the common[Pg 464] furnace, is composed of the ash-hole HIKL, the fire-place KLMN, the laboratory MNOP, and the dome RRSS, with its funnel or chimney TTVV; and to this last several additional tubes may be adapted, according to the nature of the different experiments. The retort A is placed in the division called the laboratory, and supported by two bars of iron which run across the furnace, and its beak comes out at a round hole in the side of the furnace, one half of which is cut in the piece called the laboratory, and the other in the dome. In most of the ready made reverberatory furnaces which are sold by the potters at Paris, the openings both above and below are too small: These do not allow a sufficient volume of air to pass through; hence, as the quantity of charcoal consumed, or, what is much the same thing, the quantity of caloric disengaged, is nearly in proportion to the quantity of air which passes through the furnace, these furnaces do not produce a sufficient effect in a great number of experiments. To remedy this defect, there ought to be two openings GG to the ash-hole; one of these is shut up when only a moderate fire is required; and both are kept open when the strongest power of the furnace is to be exerted. The opening of the dome SS ought likewise to be considerably larger than is usually made.[Pg 465]

It is of great importance not to employ retorts of too large size in proportion to the furnace, as a sufficient space ought always to be allowed for the passage of the air between the sides of the furnace and the vessel. The retort A in the figure is too small for the size of the furnace, yet I find it more easy to point out the error than to correct it. The intention of the dome is to oblige the flame and heat to surround and strike back or reverberate upon every part of the retort, whence the furnace gets the name of reverberatory. Without this circumstance the retort would only be heated in its bottom, the vapours raised from the contained substance would condense in the upper part, and a continual cohabitation would take place without any thing passing over into the receiver, but, by means of the dome, the retort is equally heated in every part, and the vapours being forced out, can only condense in the neck of the retort, or in the recipient.

To prevent the bottom of the retort from being either heated or coolled too suddenly, it is sometimes placed in a small sand-bath of baked clay, standing upon the cross bars of the furnace. Likewise, in many operations, the retorts are coated over with lutes, some of which are intended to preserve them from the too sudden influence of heat or of cold, while others are for sustaining the glass, or forming a kind of second[Pg 466] retort, which supports the glass one during operations wherein the strength of the fire might soften it. The former is made of brick-clay with a little cow's hair beat up alongst with it, into a paste or mortar, and spread over the glass or stone retorts. The latter is made of pure clay and pounded stone-ware mixed together, and used in the same manner. This dries and hardens by the fire, so as to form a true supplementary retort capable of retaining the materials, if the glass retort below should crack or soften. But, in experiments which are intended for collecting gasses, this lute, being porous, is of no manner of use.

In a great many experiments wherein very violent fire is not required, the reverberatory furnace may be used as a melting one, by leaving out the piece called the laboratory, and placing the dome immediately upon the fire-place, as represented Pl. XIII. Fig. 3. The furnace represented in Fig. 4. is very convenient for fusions; it is composed of the fire-place and ash-hole ABD, without a door, and having a hole E, which receives the muzzle of a pair of bellows strongly luted on, and the dome ABGH, which ought to be rather lower than is represented in the figure. This furnace is not capable of producing a very strong heat, but is sufficient for ordinary operations, and may be readily moved to any part of the laboratory[Pg 467] where it is wanted. Though these particular furnaces are very convenient, every laboratory must be provided with a forge furnace, having a good pair of bellows, or, what is more necessary, a powerful melting furnace. I shall describe the one I use, with the principles upon which it is constructed.

The air circulates in a furnace in consequence of being heated in its passage through the burning coals; it dilates, and, becoming lighter than the surrounding air, is forced to rise upwards by the pressure of the lateral columns of air, and is replaced by fresh air from all sides, especially from below. This circulation of air even takes place when coals are burnt in a common chaffing dish; but we can readily conceive, that, in a furnace open on all sides, the mass of air which passes, all other circumstances being equal, cannot be so great as when it is obliged to pass through a furnace in the shape of a hollow tower, like most of the chemical furnaces, and consequently, that the combustion must be more rapid in a furnace of this latter construction. Suppose, for instance, the furnace ABCDEF open above, and filled with burning coals, the force with which the air passes through the coals will be in proportion to the difference between the specific gravity of two columns equal to AC, the one of cold air without, and the other of heated air within the furnace.[Pg 468] There must be some heated air above the opening AB, and the superior levity of this ought likewise to be taken into consideration; but, as this portion is continually coolled and carried off by the external air, it cannot produce any great effect.

But, if we add to this furnace a large hollow tube GHAB of the same diameter, which preserves the air which has been heated by the burning coals from being coolled and dispersed by the surrounding air, the difference of specific gravity which causes the circulation will then be between two columns equal to GC. Hence, if GC be three times the length of AC, the circulation will have treble force. This is upon the supposition that the air in GHCD is as much heated as what is contained in ABCD, which is not strictly the case, because the heat must decrease between AB and GH; but, as the air in GHAB is much warmer than the external air, it follows, that the addition of the tube must increase the rapidity of the stream of air, that a larger quantity must pass through the coals, and consequently that a greater degree of combustion must take place.

We must not, however, conclude from these principles, that the length of this tube ought to be indefinitely prolonged; for, since the heat of the air gradually diminishes in passing from AB to GH, even from the contact of the sides of the[Pg 469] tube, if the tube were prolonged to a certain degree, we would at last come to a point where the specific gravity of the included air would be equal to the air without; and, in this case, as the cool air would no longer tend to rise upwards, it would become a gravitating mass, resisting the ascension of the air below. Besides, as this air, which has served for combustion, is necessarily mixed with carbonic acid gas, which is considerably heavier than common air, if the tube were made long enough, the air might at last approach so near to the temperature of the external air as even to gravitate downwards; hence we must conclude, that the length of the tube added to a furnace must have some limit beyond which it weakens, instead of strengthening the force of the fire.

From these reflections it follows, that the first foot of tube added to a furnace produces more effect than the sixth, and the sixth more than the tenth; but we have no data to ascertain at what height we ought to stop. This limit of useful addition is so much the farther in proportion as the materials of the tube are weaker conductors of heat, because the air will thereby be so much less coolled; hence baked earth is much to be preferred to plate iron. It would be even of consequence to make the tube double, and to fill the interval with rammed charcoal, which is one of the worst conductors of heat[Pg 470] known; by this the refrigeration of the air will be retarded, and the rapidity of the stream of air consequently increased; and, by this means, the tube may be made so much the longer.

As the fire-place is the hottest part of a furnace, and the part where the air is most dilated in its passage, this part ought to be made with a considerable widening or belly. This is the more necessary, as it is intended to contain the charcoal and crucible, as well as for the passage of the air which supports, or rather produces the combustion; hence we only allow the interstices between the coals for the passage of the air.

From these principles my melting furnace is constructed, which I believe is at least equal in power to any hitherto made, though I by no means pretend that it possesses the greatest possible intensity that can be produced in chemical furnaces. The augmentation of the volume of air produced during its passage through a melting furnace not being hitherto ascertained from experiment, we are still unacquainted with the proportions which should exist between the inferior and superior apertures, and the absolute size of which these openings should be made is still less understood; hence data are wanting by which to proceed upon principle, and we can only accomplish the end in view by repeated trials.[Pg 471]

This furnace, which, according to the above stated rules, is in form of an eliptical spheroid, is represented Pl. XIII. Fig. 6. ABCD; it is cut off at the two ends by two plains, which pass, perpendicular to the axis, through the foci of the elipse. From this shape it is capable of containing a considerable quantity of charcoal, while it leaves sufficient space in the intervals for the passage of the air. That no obstacle may oppose the free access of external air, it is perfectly open below, after the model of Mr Macquer's melting furnace, and stands upon an iron tripod. The grate is made of flat bars set on edge, and with considerable interstices. To the upper part is added a chimney, or tube, of baked earth, ABFG, about eighteen feet long, and almost half the diameter of the furnace. Though this furnace produces a greater heat than any hitherto employed by chemists, it is still susceptible of being considerably increased in power by the means already mentioned, the principal of which is to render the tube as bad a conductor of heat as possible, by making it double, and filling the interval with rammed charcoal.

When it is required to know if lead contains any mixture of gold or silver, it is heated in a strong fire in capsules of calcined bones, which are called cuppels. The lead is oxydated, becomes vitrified, and sinks into the substance of[Pg 472] the cuppel, while the gold or silver, being incapable of oxydation, remain pure. As lead will not oxydate without free access of air, this operation cannot be performed in a crucible placed in the middle of the burning coals of a furnace, because the internal air, being mostly already reduced by the combustion into azotic and carbonic acid gas, is no longer fit for the oxydation of metals. It was therefore necessary to contrive a particular apparatus, in which the metal should be at the same time exposed to the influence of violent heat, and defended from contact with air rendered incombustible by its passage through burning coals. The furnace intended for answering this double purpose is called the cuppelling or essay furnace. It is usually made of a square form, as represented Pl. XIII. Fig. 8. and 10. having an ash-hole AABB, a fire-place BBCC, a laboratory CCDD, and a dome DDEE. The muffle or small oven of baked earth GH, Fig. 9. being placed in the laboratory of the furnace upon cross bars of iron, is adjusted to the opening GG, and luted with clay softened in water. The cuppels are placed in this oven or muffle, and charcoal is conveyed into the furnace through the openings of the dome and fire-place. The external air enters through the openings of the ash-hole for supporting the combustion, and escapes by the superior opening or chimney at EE; and air is[Pg 473] admitted through the door of the muffle GG for oxydating the contained metal.

Very little reflection is sufficient to discover the erroneous principles upon which this furnace is constructed. When the opening GG is shut, the oxydation is produced slowly, and with difficulty, for want of air to carry it on; and, when this hole is open, the stream of cold air which is then admitted fixes the metal, and obstructs the process. These inconveniencies may be easily remedied, by constructing the muffle and furnace in such a manner that a stream of fresh external air should always play upon the surface of the metal, and this air should be made to pass through a pipe of clay kept continually red hot by the fire of the furnace. By this means the inside of the muffle will never be coolled, and processes will be finished in a few minutes, which at present require a considerable space of time.

Mr Sage remedies these inconveniencies in a different manner; he places the cuppel containing lead, alloyed with gold or silver, amongst the charcoal of an ordinary furnace, and covered by a small porcelain muffle; when the whole is sufficiently heated, he directs the blast of a common pair of hand-bellows upon the surface of the metal, and completes the cuppellation in this way with great ease and exactness.[Pg 474]

SECT. III.
Of increasing the Action of Fire, by using Oxygen Gas instead of Atmospheric Air.

By means of large burning glasses, such as those of Tchirnausen and Mr de Trudaine, a degree of heat is obtained somewhat greater than has hitherto been produced in chemical furnaces, or even in the ovens of furnaces used for baking hard porcelain. But these instruments are extremely expensive, and do not even produce heat sufficient to melt crude platina; so that their advantages are by no means sufficient to compensate for the difficulty of procuring, and even of using them. Concave mirrors produce somewhat more effect than burning glasses of the same diameter, as is proved by the experiments of Messrs Macquer and Beaumé with the speculum of the Abbé Bouriot; but, as the direction of the reflected rays is necessarily from below upwards, the substance to be operated upon must be placed in the air without any support, which renders most chemical experiments impossible to be performed with this instrument.[Pg 475]

For these reasons, I first endeavoured to employ oxygen gas for combustion, by filling large bladders with it, and making it pass through a tube capable of being shut by a stop-cock; and in this way I succeeded in causing it to support the combustion of lighted charcoal. The intensity of the heat produced, even in my first attempt, was so great as readily to melt a small quantity of crude platina. To the success of this attempt is owing the idea of the gazometer, described p. 308. *et seq.* which I substituted instead of the bladders; and, as we can give the oxygen gas any necessary degree of pressure, we can with this instrument keep up a continued stream, and give it even a very considerable force.

The only apparatus necessary for experiments of this kind consists of a small table ABCD, Pl. XII. Fig. 15, with a hole F, through which passes a tube of copper or silver, ending in a very small opening at G, and capable of being opened or shut by the stop-cock H. This tube is continued below the table at *l m n o*, and is connected with the interior cavity of the gazometer. When we mean to operate, a hole of a few lines deep must be made with a chizel in a piece of charcoal, into which the substance to be treated is laid; the charcoal is set on fire by means of a candle and blow-pipe, after which it is exposed[Pg 476] to a rapid stream of oxygen gas from the extremity G of the tube FG.

This manner of operating can only be used with such bodies as can be placed, without inconvenience, in contact with charcoal, such as metals, simple earths, &c. But, for bodies whose elements have affinity to charcoal, and which are consequently decomposed by that substance, such as sulphats, phosphats, and most of the neutral salts, metallic glasses, enamels, &c. we must use a lamp, and make the stream of oxygen gas pass through its flame. For this purpose, we use the elbowed blow-pipe ST, instead of the bent one FG, employed with charcoal. The heat produced in this second manner is by no means so intense as in the former way, and is very difficultly made to melt platina. In this manner of operating with the lamp, the substances are placed in cuppels of calcined bones, or little cups of porcelain, or even in metallic dishes. If these last are sufficiently large, they do not melt, because, metals being good conductors of heat, the caloric spreads rapidly through the whole mass, so that none of its parts are very much heated.

In the Memoirs of the Academy for 1782, p. 476. and for 1783, p. 573. the series of experiments I have made with this apparatus may be seen at large. The following are some of the principal results.[Pg 477]

1. Rock cristal, or pure silicious earth, is infusible, but becomes capable of being softened or fused when mixed with other substances.

2. Lime, magnesia, and barytes, are infusible, either when alone, or when combined together; but, especially lime, they assist the fusion of every other body.

3. Argill, or pure base of alum, is completely fusible *per se* into a very hard opake vitreous substance, which scratches glass like the precious stones.

4. All the compound earths and stones are readily fused into a brownish glass.

5. All the saline substances, even fixed alkali, are volatilized in a few seconds.

6. Gold, silver, and probably platina, are slowly volatilized without any particular phenomenon.

7. All other metallic substances, except mercury, become oxydated, though placed upon charcoal, and burn with different coloured flames, and at last dissipate altogether.

8. The metallic oxyds likewise all burn with flames. This seems to form a distinctive character for these substances, and even leads me to believe, as was suspected by Bergman, that barytes is a metallic oxyd, though we have not hitherto been able to obtain the metal in its pure or reguline state.[Pg 478]

9. Some of the precious stones, as rubies, are capable of being softened and soldered together, without injuring their colour, or even diminishing their weights. The hyacinth, tho' almost equally fixed with the ruby, loses its colour very readily. The Saxon and Brasilian topaz, and the Brasilian ruby, lose their colour very quickly, and lose about a fifth of their weight, leaving a white earth, resembling white quartz, or unglazed china. The emerald, chrysolite, and garnet, are almost instantly melted into an opake and coloured glass.

10. The diamond presents a property peculiar to itself; it burns in the same manner with combustible bodies, and is entirely dissipated.

There is yet another manner of employing oxygen gas for considerably increasing the force of fire, by using it to blow a furnace. Mr Achard first conceived this idea; but the process he employed, by which he thought to dephlogisticate, as it is called, atmospheric air, or to deprive it of azotic gas, is absolutely unsatisfactory. I propose to construct a very simple furnace, for this purpose, of very refractory earth, similar to the one represented Pl.

XIII. Fig. 4. but smaller in all its dimensions. It is to have two openings, as at E, through one of which the nozle of a pair of bellows is to pass, by which the heat is to be raised as high as possible with common air; after which, the[Pg 479]stream of common air from the bellows being suddenly stopt, oxygen gas is to be admitted by a tube, at the other opening, communicating with a gazometer having the pressure of four or five inches of water. I can in this manner unite the oxygen gas from several gazometers, so as to make eight or nine cubical feet of gas pass through the furnace; and in this way I expect to produce a heat greatly more intense than any hitherto known. The upper orifice of the furnace must be carefully made of considerable dimensions, that the caloric produced may have free issue, lest the too sudden expansion of that highly elastic fluid should produce a dangerous explosion.

FINIS.

[Pg 481]

APPENDIX.

No. I.

TABLE *for Converting Lines, or Twelfth Parts of an Inch, and Fractions of Lines, into Decimal Fractions of the Inch.*

Twelfth Parts of a Line.	Decimal Fractions.	ines.	Decimal Fractions.
1	0.00694		0.08333
2	0.01389	:	0.16667
3	0.02083	:	0.25000
4	0.02778	,	0.33333
5	0.03472	!	0.41667
6	0.04167	(	0.50000
7	0.04861		0.58333
8	0.05556	;	0.66667
9	0.06250	'	0.75000
10	0.06944	0	0.83333
11	0.07639	1	0.91667
12	0.08333	2	1.00000

[Pg 482]

No. II.

TABLE *for Converting the Observed Heighths of Water in the Jars of the Pneumato-Chemical Apparatus, expressed in Inches and Decimals, into Corresponding Heighths of Mercury.*

Water.	Mercury.	Water.	Mercury.

Ounce measures	French cubical inches	Ounce measures	French cubical inches
1	.0737	4.	.29480
2	.1474	5.	.36851
3	.2201	6.	.44221
4	.2948	7.	.51591
5	.3685	8.	.58961
6	.4422	9.	.66332
7	.5159	10.	.73702
8	.5896	11.	.81072
9	.6633	12.	.88442
10.	.7370	13.	.96812
20.	1.4740	14.	1.04182
30.	2.2010	15.	1.11525

[Pg 483]

No. III.
TABLE *for Converting the Ounce Measures used by Dr Priestly into French and English Cubical Inches.*

Ounce measures.	French cubical inches.	English cubical inches.
1	1.567	1.898
2	3.134	3.796
3	4.701	5.694
4	6.268	7.592
5	7.835	9.490
6	9.402	11.388
7	10.969	13.286

8	12.536	15.184
9	14.103	17.082
10	15.670	18.980
20	31.340	37.960
30	47.010	56.940
40	62.680	75.920
50	78.350	94.900
60	94.020	113.880
70	109.690	132.860
80	125.360	151.840
90	141.030	170.820
100	156.700	189.800
1000	1567.000	1898.000

[Pg 484]

No. IV. ADDITIONAL.

TABLE *for Reducing the Degrees of Reaumeur's Thermometer into its corresponding Degrees of Fahrenheit's Scale.*

F	.	F	.	F	.	F	.
2	1	7 9.25	1	1 24.25	1	1 69.25	1
4.25	2	8 1.5	2	1 26.5	2	1 71.5	2
6.5	3	8 3.75	3	1 28.75	3	1 73.75	3
8.75	4	8 6	4	1 31	4	1 76.	4
1	5	8 8.25	5	1 33.25	5	1 78.25	5
3.25	6	9 0.5	6	1 35.5	6	1 80.5	6
5.5	7	9 2.75	7	1 37.75	7	1 82.75	7
7.75	8	9 5	8	1 40	8	1 85	8

			⁹ 7.25	¹	42.25	¹	87.25
0	9						
			⁹ 9.5	¹	44.5	¹	89.5
	2.25	0		0		0	
0	4.5	1	¹01.75	1	¹46.75	1	¹91.75
1	6.75	2	¹04	2	¹49	2	¹94.
2	9	3	¹06.25	3	¹51.25	3	¹96.25
3	1.25	4	¹08.5	4	¹53.5	4	¹98.5
4	3.5	5	¹10.75	5	¹55.75	5	²00.75
5	5.75	6	¹13	6	¹58	6	03
6	8	7	¹15.25	7	¹60.25	7	05.25
7	0.25	8	¹17.5	8	¹62.5	8	07.5
8	2.5	9	¹19.75	9	¹64.75	9	09.75
9	4.75	0	¹22	0	¹67	0	12
0	7						

Note—Any degree, either higher or lower, than what is contained in the above Table, may be at any time converted, by remembering that one degree of Reaumur's scale is equal to 2.25° of Fahrenheit; or it may be done without the Table by the following formula, R × 9 / 4 + 32 = F; that is, multiply the degree of Reaumur by 9, divide the product by 4, to the quotient add 32, and the sum is the degree of Fahrenheit.—E.[Pg 485]

No. V. Additional.

RULES *for converting French Weights and Measures into correspondent English Denominations*[62].

§ 1. *Weights.*

The Paris pound, poids de mark of Charlemagne, contains 9216 Paris grains; it is divided into 16 ounces, each ounce into 8 gros, and each gros into 72 grains. It is equal to 7561 English Troy grains.

The English Troy pound of 12 ounces contains 5760 English Troy grains, and is equal to 7021 Paris grains.

The English averdupois pound of 16 ounces contains 7000 English Troy grains, and is equal to 8538 Paris grains.

To reduce Paris *grs.* to English   1.

Troy *grs.* divide by 2189

To reduce English Troy *grs.* to Paris *grs.* multiply by

To reduce Paris ounces to English Troy, divide by

To reduce English Troy ounces to Paris, multiply by 1.015734

[Pg 486]
Or the conversion may be made by means of the following Tables.

I. *To reduce French to English Troy Weight.*

The Paris pound	7561	}
The ounce	472.5625	} English.
The gros	59.0703	} Troy.
The grain	.8194	} Grains.

II. *To Reduce English Troy to Paris Weight.*

The English Troy pound of 12 ounces	7021.	}
The Troy ounce	585.0830	}
The dram of 60 grs.	73.1353	} Paris
The penny weight, or denier, of 24 grs.	29.2540	} grains.
The scruple, of 20 grs.	24.3784	}

III. *To Reduce English Averdupois to Paris Weight.*

The averdupois pound of 16 ounces, or 7000 Troy grains.	8538.	} Paris
The ounce	533.6250	} grains.

[Pg 487]
§ 2. *Long and Cubical Measures.*

To reduce Paris feet or inches into English, multiply by 1.065977

English feet or inches into Paris, divide by

To reduce Paris cubic feet or inches to English, multiply by 1.211278

English cubic feet or inches to Paris, divide by

Or by means of the following tables:

IV. To Reduce Paris Long Measure to English.

The Paris royal foot of 12 inches	1 2.7977	} English
The inch	1 .0659	}
The line, or 1/12 of an inch	.0888	} inches.
The 1/12 of a line	.0074	}

V. To Reduce English Long Measure to French.

The English foot	1 1.2596	}
The inch	.9383	}
The 1/8 of an inch	.1173	} Paris inches.
The 1/10	.0938	}
The line, or 1/12	.0782	}

[Pg 488]

VI. To Reduce French Cube Measure to English.

The Paris cube foot	1.211278	{2093.088384	}
The cubic inch	.000700	English cubical feet, or {1.211278	} inches.

VII. To Reduce English Cube Measure to French.

The English cube foot, or 1728 cubical inches	14 27.4864	} French
The cubical inch	.8260	} cubical
The cube tenth	.0008	} inches.

§ 3. Measure of Capacity.

The Paris pint contains 58.145[63] English cubical inches, and the English wine pint contains 28.85 cubical inches; or, the Paris pint contains[Pg 489] 2.01508 English pints, and the English pint contains .49617 Paris pints; hence,

To reduce the Paris pint to the English, multiply by 2.01508.

To reduce the English pint to the Paris, divide by

[Pg 490]

No. VI.
TABLE *of the Weights of the different Gasses, at 28 French inches, or 29.84 English inches barometrical pressure, and at 10° (54.5°) of temperature, expressed in English measure and English Troy weight.*

Names of the Gasses.	Weight of a cubical inch.		Weight of a cubical foot.		
(A)	qrs.	z.	r.	s.	qr
Atmospheric air	.32112				15
Azotic gas	.30064	.5			39
Oxygen gas	.34211				51
Hydrogen gas	.02394	.26			41
Carbonic acid gas	.44108				41
(B)					
Nitrous gas	.37000				39
Ammoniacal gas	.18515	.73			19
Sulphurous acid gas	.71580				38

[Note A: These five were ascertained by Mr Lavoisier himself.—E.]
[Note B: The last three are inserted by Mr Lavoisier upon the authority of Mr Kirwan.—E.][Pg 491]

No. VII.
Tables of the Specific Gravities of different bodies.
§ 1. *Metallic Substances.*
GOLD.

Pure gold of 24 carats melted but not hammered	1 9.2581
The same hammered	1 9.3617

186

Gold of the Parisian standard, 22 carats fine, not hammered(A)	$\frac{1}{7.4863}$
The same hammered	$\frac{1}{7.5894}$
Gold of the standard of French coin, 21-22/32 carats fine, not hammered	$\frac{1}{7.4022}$
The same coined	$\frac{1}{7.6474}$
Gold of the French trinket standard, 20 carats fine, not hammered	$\frac{1}{5.7090}$
The same hammered	$\frac{1}{5.7746}$

[Note A: The same with Sterling.]

SILVER.

Pure or virgin silver, 24 deniers, not hammered	$\frac{1}{0.4743}$
The same hammered	$\frac{1}{0.5107}$
Silver of the Paris standard, 11 deniers 10 grains fine, not hammered(B)	$\frac{1}{0.1752}$
The same hammered	$\frac{1}{0.3765}$

[Pg 492]

Silver, standard of French coin, 10 deniers 21 grains fine, not hammered	$\frac{1}{0.0476}$
The same coined	$\frac{1}{0.4077}$

[Note B: This is 10 *grs.* finer than Sterling.]

PLATINA.

Crude platina in grains	$\frac{1}{5.6017}$
The same, after being treated with muriatic acid	$\frac{1}{6.7521}$
Purified platina, not hammered	$\frac{1}{9.5000}$
The same hammered	$\frac{2}{0.3366}$
The same drawn into wire	$\frac{2}{1.0417}$

The same passed through rollers	2.0690²

COPPER AND BRASS.

Copper, not hammered	7.7880
The same wire drawn	8.8785
Brass, not hammered	8.3958
The same wire drawn	8.5441

IRON AND STEEL.

Cast iron	7.2070
Bar iron, either screwed or not	7.7880
Steel neither tempered nor screwed	7.8331
Steel screwed but not tempered	7.8404
Steel tempered and screwed	7.8180
Steel tempered and not screwed	7.8163

[Pg 493]

TIN.

Pure tin from Cornwall melted and not screwed	7.2914
The same screwed	7.2994
Malacca tin, not screwed	7.2963
The same screwed	7.3065
Molten lead	11.3523
Molten zinc	7.1908
Molten bismuth	9

	.8227
Molten cobalt	.8119⁷
Molten arsenic	.7633⁵
Molten nickel	.8070⁷
Molten antimony	.7021⁶
Crude antimony	.0643⁴
Glass of antimony	.9464⁴
Molybdena	.7385⁴
Tungstein	.0665⁶
Mercury	3.5681¹

§ 2. *Precious Stones.*

White Oriental diamond	.5212³
Rose-coloured Oriental ditto	.5310³
Oriental ruby	.2833⁴
Spinell ditto	.7600³
Ballas ditto	.6458³
Brasillian ditto	.5311³
Oriental topas	.0106⁴

[Pg 494]

Ditto Pistachio ditto	.0615⁴
Brasillian ditto	3

		.5365
Saxon topas		3.5640
Ditto white ditto		3.5535
Oriental saphir		3.9941
Ditto white ditto		3.9911
Saphir of Puy		4.0769
Ditto of Brasil		3.1307
Girasol		4.0000
Ceylon jargon		4.4161
Hyacinth		3.6873
Vermillion		4.2299
Bohemian garnet		4.1888
Dodecahedral ditto		4.0627
Syrian ditto		4.0000
Volcanic ditto, with 24 sides		2.4684
Peruvian emerald		2.7755
Crysolite of the jewellers		2.7821
Ditto of Brasil		2.6923
Beryl, or Oriental aqua marine		3.5489
Occidental aqua		2

marine	.7227

§ 3. *Silicious Stones.*

Pure rock cristal of Madagascar	2.6530
Ditto of Brasil	2.6526
Ditto of Europe, or gelatinous	2.6548
Cristallized quartz	2.6546
Amorphous ditto	2.6471

[Pg 495]

Oriental agate	2.5901
Agate onyx	2.6375
Transparent calcedony	2.6640
Carnelian	2.6137
Sardonyx	2.6025
Prase	2.5805
Onyx pebble	2.6644
Pebble of Rennes	2.6538
White jade	2.9502
Green jade	2.9660
Red jasper	2.6612
Brown ditto	2.6911
Yellow ditto	2

Violet ditto	2.7101
Gray ditto	2.7111
Jasponyx	2.7640
Black prismatic hexahedral schorl	2.8160
Black spary ditto	3.3852
Black amorphous schorl, called antique basaltes	3.3852
Paving stone	2.9225
Grind stone	2.4158
Cutler's stone	2.1429
Fountainbleau stone	2.1113
Scyth stone of Auvergne	2.5616
Ditto of Lorrain	2.5638
Mill stone	2.5298
White flint	2.4835
Blackish ditto	2.5941
	2.5817

[Pg 496]

§ 4. *Various Stones, &c.*

Opake green Italian serpentine, or gabro of the Florentines	2.4295
Coarse Briancon chalk	2.7274
Spanish chalk	2

		2.7902
Foliated lapis ollaris of Dauphiny		2.7687
Ditto ditto from Sweden		2.8531
Muscovy talc		2.7917
Black mica		2.9004
Common schistus or slate		2.6718
New slate		2.8535
White rasor hone		2.8763
Black and white hone		3.1311
Rhombic or Iceland cristal		2.7151
Pyramidal calcareous spar		2.7141
Oriental or white antique alabaster		2.7302
Green Campan marble		2.7417
Red Campan marble		2.7242
White Carara marble		2.7168
White Parian marble		2.8376
Various kinds of calcareous stones	from	1.3864
used in France for building.	to	2.3902
Heavy spar		4.4300
White fluor		3

	3.1555
Red ditto	3.1911
Green ditto	3.1817
Blue ditto	3.1688
Violet ditto	3.1757

[Pg 497]

Red scintilant zeolite from Edelfors	2.4868
White scintilant zeolite	2.0739
Cristallized zeolite	2.0833
Black pitch stone	2.0499
Yellow pitch stone	2.0860
Red ditto	2.6695
Blackish ditto	2.3191
Red porphyry	2.7651
Ditto of Dauphiny	2.7033
Green serpentine	2.8960
Black ditto of Dauphiny, called variolite	2.9339
Green ditto from Dauphiny	2.9883
Ophites	2.9722
Granitello	3.0626

Red Egyptian granite	2.6541
Beautiful red granite	2.7609
Granite of Girardmas	2.7163
Pumice stone	.9145
Lapis obsidianus	2.3480
Pierre de Volvic	2.3205
Touch stone	2.4153
Basaltes from Giants Causeway	2.8642
Ditto prismatic from Auvergne	2.4153
Glass gall	2.8548
Bottle glass	2.7325
Green glass	2.6423
White glass	2.8922
St Gobin cristal	2.4882
Flint glass	3.3293
Borax glass	2.6070

[Pg 498]

Seves porcelain	2.1457
Limoges ditto	2.3410
China ditto	2

	.3847
Native sulphur	2.0332
Melted sulphur	1.9907
Hard peat	1.3290
Ambergrease	.9263
Yellow transparent amber	1.0780

§ 5. *Liquids.*

Distilled water	1.0000
Rain water	1.0000
Filtered water of the Seine	1.00015
Arcueil water	1.00046
Avray water	1.00043
Sea water	1.0263
Water of the Dead Sea	1.2403
Burgundy wine	.9915
Bourdeaux ditto	.9939
Malmsey Madeira	1.0382
Red beer	1.0338
White ditto	1.0231
Cyder	1.0181

Highly rectified alkohol				.8293
Common spirits of wine				.8371

[Pg 499]

Alkohol	Water	part.	
15 pts.	1		.8527
14	2		.8674
13	3		.8815
12	4		.8947
11	5		.9075
10	6		.9199
9	7		.9317
8	8		.9427
7	9		.9519
6	10		.9594
5	11		.9674
4	12		.9733
3	13		.9791
2	14		.9852
1	15		.9919

| Sulphuric ether | | .7394 |
| Nitric ether | | . |

		.9088
Muriatic ether		.7298
Acetic ether		.8664
Sulphuric acid		1.8409
Nitric ditto		1.2715
Muriatic ditto		1.1940
Red acetous ditto		1.0251
White acetous ditto		1.0135
Distilled ditto ditto		1.0095
Acetic ditto		1.0626
Formic ditto		.9942
Solution of caustic ammoniac,	or volatil alkali fluor	.8970

[Pg 500]

Essential or volatile oil	of turpentine	.8697
Liquid turpentine		.9910
Volatile oil of lavender		.8938
Volatile oil of cloves		1.0363
Volatile oil of cinnamon		1.0439
Oil of olives		.9153
Oil of sweet almonds		.9170

Lintseed oil	.9403
Oil of poppy seed	.9288
Oil of beech mast	.9176
Whale oil	.9233
Womans milk	1.0203
Mares milk	1.0346
Ass milk	1.0355
Goats milk	1.0341
Ewe milk	1.0409
Cows milk	1.0324
Cow whey	1.0193
Human urine	1.0106

§ 6. Resins and Gums

Common yellow or white rosin	1.0727
Arcanson	1.0857
Galipot(A)	1.0819
Baras(A)	1.0441

[Pg 501]

Sandarac	1.0920
Mastic	1.0742

Storax	1.1098
Opake copal	1.1398
Transparent ditto	1.0452
Madagascar ditto	1.0600
Chinese ditto	1.0628
Elemi	1.0182
Oriental anime	1.0284
Occidental ditto	1.0426
Labdanum	1.1862
Ditto *in tortis*	2.4933
Resin of guaiac	1.2289
Ditto of jallap	1.2185
Dragons blood	1.2045
Gum lac	1.1390
Tacamahaca	1.0463
Benzoin	1.0924
Alouchi(B)	1.0604
Caragna(C)	1.1244
Elastic gum	.9335

Camphor	.9887
Gum ammoniac	1.2071
Sagapenum	1.2008

[Pg 502]

Ivy gum(D)	1.2948
Gamboge	1.2216
Euphorbium	1.1244
Olibanum	1.1732
Myrrh	1.3600
Bdellium	1.3717
Aleppo Scamony	1.2354
Smyrna ditto	1.2743
Galbanum	1.2120
Assafoetida	1.3275
Sarcocolla	1.2684
Opoponax	1.6226
Cherry tree gum	1.4817
Gum Arabic	1.4523
Tragacanth	1.3161
Basora gum	1

	.4346
Acajou gum(E)	1.4456
Monbain gum(F)	1.4206
Inspissated juice of liquorice	1.7228
——— Acacia	1.5153
——— Areca	1.4573
Terra Japonica	1.3980
Hepatic aloes	1.3586
Socotrine aloes	1.3795
Inspissated juice of St John's wort	1.5263

[Pg 503]

Opium	1.3366
Indigo	.7690
Arnotto	.5956
Yellow wax	.9648
White ditto	.9686
Ouarouchi ditto(G)	.8970
Cacao butter	.8916
Spermaceti	.9433
Beef fat	.9232

Veal fat	9342
Mutton fat	9235
Tallow	9419
Hoggs fat	9368
Lard	9478
Butter	9423

[Note A: Resinous juices extracted in France from the Pine. *Vide Bomare's Dict.*]
[Note B: Odoriferous gum from the tree which produces the Cortex Winteranus.*Bomare.*]
[Note C: Resin of the tree called in Mexico Caragna, or Tree of Madness. *Ibid.*]
[Note D: Extracted in Persia and the warm countries from Hedera terrestris.—*Bomare.*]
[Note E: From a Brasilian tree of this name.—*Ibid.*]
[Note F: From a tree of this name.—*Ibid.*]
[Note G: The produce of the Tallow Tree of Guayana. *Vide Bomare's Dict.*]

§ 7. *Woods.*

Heart of oak 60 years old	1.1700
Cork	2400
Elm trunk	6710
Ash ditto	8450
Beech	8520
Alder	8000
Maple	7550
Walnut	6710
Willow	5850
Linden	

		6040
[Pg 504]		
	Male fir	5500
	Female ditto	4980
	Poplar	3830
ditto	White Spanish	5294
	Apple tree	7930
	Pear tree	6610
	Quince tree	7050
	Medlar	9440
	Plumb tree	7850
	Olive wood	9270
	Cherry tree	7150
	Filbert tree	6000
	French box	9120
	Dutch ditto	1.3280
	Dutch yew	7880
	Spanish ditto	8070
	Spanish cypress	6440
	American cedar	5608

Pomgranate tree	1.3540
Spanish mulberry tree	.8970
Lignum vitae	1.3330
Orange tree	.7050

Note—The numbers in the above Table, if the Decimal point be carried three figures farther to the right hand, nearly express the absolute weight of an English cube foot of each substance in averdupois ounces. See No. VIII. of the Appendix.—E.[Pg 505]

No. VIII. ADDITIONAL.

RULES *for Calculating the Absolute Gravity in English Troy Weight of a Cubic Foot and Inch, English Measure, of any Substance whose Specific Gravity is known*[64].

In 1696, Mr Everard, balance-maker to the Exchequer, weighed before the Commissioners of the House of Commons 2145.6 cubical inches, by the Exchequer standard foot, of distilled water, at the temperature of 55° of Fahrenheit, and found it to weigh 1131 oz. 14 dts. Troy, of the Exchequer standard. The beam turned with 6 grs. when loaded with 30 pounds in each scale. Hence, supposing the pound averdupois to weigh 7000 grs. Troy, a cubic foot of water weighs 62-1/2 pounds averdupois, or 1000 ounces averdupois, wanting 106 grains Troy. And hence, if the specific gravity of water be called 1000, the proportional specific gravities of all other bodies will nearly express the number of averdupois ounces in a cubic foot. Or more accurately, supposing the specific gravity of water expressed by 1. and of all other bodies in proportional numbers, as the[Pg 506] cubic foot of water weighs, at the above temperature, exactly 437489.4 grains Troy, and the cubic inch of water 253.175 grains, the absolute weight of a cubical foot or inch of any body in Troy grains may be found by multiplying their specific gravity by either of the above numbers respectively.

By Everard's experiment, and the proportions of the English and French foot, as established by the Royal Society and French Academy of Sciences, the following numbers are ascertained.

Paris grains in a Paris cube foot of water	6 45511
English grains in a Paris cube foot of water	5 29922
Paris grains in an English cube foot of water	5 33247
English grains in an English cube foot of water	4 37489.4
English grains in an English cube inch of	2

water 53.175

By an experiment of Picard with the measure and weight of the Chatelet, the Paris cube foot of water contains of Paris grains 641326

By one of Du Hamel, made with great care 641376

By Homberg 641666

[Pg 507]
These show some uncertainty in measures or in weights; but the above computation from Everard's experiment may be relied on, because the comparison of the foot of England with that of France was made by the joint labours of the Royal Society of London and the French Academy of Sciences: It agrees likewise very nearly with the weight assigned by Mr Lavoisier, 70 Paris pounds to the cubical foot of water.[Pg 508]

No. IX.

TABLES *for Converting Ounces, Drams, and Grains, Troy, into Decimals of the Troy Pound of 12 Ounces, and for Converting Decimals of the Pound Troy into Ounces, &c.*

I. For Grains.

Grains	= Pound.
1	.0001736
2	.0003472
3	.0005208
4	.0006944
5	.0008681
6	.0010417
7	.00

	12153	
8	13889	.00
9	15625	.00
10	17361	.00
20	34722	.00
30	52083	.00
40	69444	.00
50	86806	.00
60	04167	.01
70	21528	.01
80	38889	.01
90	56250	.01
100	73611	.01
200	74222	.03
300	20833	.05
400	94444	.06
500	68055	.08
600	41666	.10

00	7 15277	.12
00	8 88888	.13
00	9 62499	.15
000	1 36110	.17
000	2 72220	.34
000	3 08330	.52
000	4 44440	.69
000	5 80550	.86
000	6 418660	1.0
000	7 152770	1.2
000	8 888880	1.3
000	9 624990	1.5

[Pg 509]

II. *For Drams.*

rams	= Pound.
1	.0104167
2	.0208333
3	.0312500
4	.0416667
5	.0520833

6   .0
    625000

7   .0
    729167

8   .0
    833333

III. *For Ounces.*

Ounces	= Pounds.
1	.08333333
2	.1666667
3	.2500000
4	.3333333
5	.4166667
6	.5000000
7	.5833333
8	.6666667
9	.7500000
10	.8333333
11	.9166667
12	1.0000000

[Pg 510]

IV. *Decimals of the Pound into Ounces, &c.*

Tenth parts.

lb. =	*l* z.	r.	r.
0			

.1	0	6
.2	0	2
.3	0	8
.4	0	4
.5	0	
.6	0	6
.7	0	2
.8	0	8
.9	0	4

Hundredth parts.

.01	0	7.6
.02	0	5.2
.03	0	2.8
.04	0	0.4
.05	0	8.0
.06	0	5.6
.07	0	3.2
.08	0	0.8
.09	0	8.4

Thousandths.

.001   0.76

.002   0.152 → 1.52

.003   0 7.28

.004   0 3.04

.005   0 8.80

1 ib. = rs.

.006   0 4.56

.007   0 0.32

.008   0 6.08

.009   0 1.84

Ten thousandth parts.

.0001   0.576

.0002   0.152

.0003   0.728

.0004   0.304

.0005   0.880

.0006   0.456

.0007   0.032

.0008   0.608

0

.0009	.184

parts. Hundred thousandth

0.00001	0.052
0.00002	0.115
0.00003	.173
0.00004	.230
0.00005	.288
0.00006	.346
0.00007	.403
0.00008	.461
0.00009	.518

No. X.

TABLE *of the English Cubical Inches and Decimals corresponding to a determinate Troy Weight of Distilled Water at the Temperature of 55°, calculated from Everard's experiment.*

For Grains.

rs.	Cubical inches.
=	.0039
	.0078
	.0118
	.0157
	.0197
	.0236
	.0275
	.0315

		.0354
0		.0394
0		.0788
0		.1182
0		.1577
0		.1971

For Drams.

rams.	Drams.	Cubical inches.
=	1	.2365
	2	.4731
	3	.7094
	4	.9463
	5	1.1829
	6	1.4195
	7	1.6561

For Ounces.

z.	Cubical inches.
=	1.8927
	3.7855
	5.6782
	7.5710
	9.4631
	11.3565
	13.2493
	15.1420

		17.0748
0		18.9276
1		20.8204

For Pounds.

lbs.	Cubical inches.
=	22.7131
	45.4263
	68.1394
	90.8525
7	113.565
8	136.278
9	158.991
1	181.705
3	204.418
0 4	227.131
0 74	1135.65
00 48	2271.31
000 488	22713.1

THE END.

FOOTNOTES:

[62] For the materials of this Article the Translator is indebted to Professor Robertson.

[63] It is said, *Belidor Archit. Hydrog.* to contain 31 *oz.* 64 *grs.* of water, which makes it 58.075 English inches; but, as there is considerable uncertainty in the determinations of the weight of the French cubical measure of water, owing to the uncertainty of the standards made use of, it is better to abide by Mr Everard's measure, which was with the Exchequer

standards, and by the proportions of the English and French foot, as established by the French Academy and Royal Society.

[64]The whole of this and the following article was communicated to the Translator by Professor Robinson.—E.

[Pg 512]

www.ingramcontent.com/pod-product-compliance
Lightning Source LLC
Chambersburg PA
CBHW051802170526
45167CB00005B/1851